Science, Ethics, and Politics

Science, Ethics, and Politics

Conversations and Investigations

Edited by
Kristen Renwick Monroe

Paradigm Publishers
Boulder • London

Published in the United States by Paradigm Publishers, 2845 Wilderness Place, Boulder, CO 80301 USA.

Paradigm Publishers is the trade name of Birkenkamp & Company, LLC,
Dean Birkenkamp, President and Publisher.

Library of Congress Cataloging-in-Publication Data for this title is available from the Library of Congress.

ISBN 978-1-59451-997-0 (paperback : alk. paper)

Printed and bound in the United States of America on acid-free paper that meets the standards of the American National Standard for Permanence of Paper for Printed Library Materials.

Designed and Typeset by Straight Creek Bookmakers.

16 15 14 13 12 1 2 3 4 5

With gratitude for the friendship and support of David Easton and Bettye Vaughen.

In loving memory of Frank Lynch.

Contents

Introduction

Kristen Renwick Monroe

Science and ethics are unusual bedfellows. Traditionally, scientists proceed from empirical observation of some phenomenon to constructing explanatory theories that are seldom normative in nature. Conversely, ethical issues conventionally fall in the domain of the philosopher or the theologian and start with first principles. Courses on ethics typically stress the classics, telling students of the works of Kant or the Utilitarians, for example, and then posing ethical dilemmas to help the student work through the process of moral choice after having been informed of the canon in ethics. In this volume, we take a different tack. We present work that utilizes the tools of science—broadly conceptualized—to elucidate ethical issues, assuming that careful scientific analysis of ethical issues can reveal new insights. We supplement this with conversations with the authors—some of the pre-eminent scientists addressing issues of ethics—to learn how they came to the study of ethics and how they analyze and think about ethical issues. This volume thus provides substantive insight into particular ethical issues—ranging from issues of torture during war to parents' obligations to children—but also reveals how a scientific approach might lead to different insights into these issues than does the more traditional approach.

All of the chapters in this volume were presented at the University of California at Irvine, at the UCI Interdisciplinary Center for the Scientific Study of Ethics and Morality. The authors come from widely diverse fields, from economics and political science to biology, logic and philosophy of science. We publish these to demonstrate the contribution of the scientific analysis of ethical issues and to encourage others to pursue such work. Each chapter is preceded by a short interview with one of the authors, to set the focus for their work. We ask authors how they came to do this particular work, what they believe their contribution has been to the debate, and what unanswered questions they hope others will focus on in later work. We tried to choose some of the most interesting original pieces presented at the Center, on topics that would have broad appeal. Space constraints meant many excellent works could not be included; perhaps some of these can appear in a later volume.

The book opens by directly addressing the issue of science and ethics. Francisco J. Ayala, the 2010 recipient of the Templeton Award for his work on science and religion, sets the tone for the volume by arguing that nothing in the world of nature is outside science's province, a universality we owe to Darwin's discovery of natural selection. Ayala begins by showing how Copernicus, Kepler, Galileo, and Newton, in the sixteenth and seventeenth centuries, ushered in a conception of the universe as matter in motion governed by natural laws. Darwin completed the Copernican revolution by extending it to the living world with his theory of natural selection, the process that explains the design of organisms. The origin of species and the exquisite features of organisms had previously been explained as special creations of an omniscient God. Darwin brought them into the domain of science. The argument-from-design to demonstrate the existence of God based on the functional design of organisms has been revived in the last two decades in the form of the so-called Intelligent Design (ID). Ayala argues that intelligent design has no scientific cogency whatsoever. Moreover, he finds that intelligent design imputes attributes of the Creator contrary to those held by many people of faith. Indeed, Ayala concludes that in view of current biological knowledge, intelligent design amounts to blasphemy.

Ayala concludes with a brief reflection about the power and limits of science. Science has been extremely successful and of great consequence for industrial and economic development and for human health. Ayala suggests science is one way of knowing but it is not the only way. Knowledge also has other sources, such as common sense, artistic and religious experience, and philosophical reflection. After establishing the claim that nothing in the world of nature is outside science's scope, Ayala further observes that a scientific view of the world is hopelessly incomplete. The significance of life and the world, questions of value and meaning, and esthetic and moral perceptions lie beyond science's scope.

Rose McDermott's chapter continues with a concern for biological science and is centrally located in a major problem that preoccupies much work on ethics and international relations: the causes and sources of conflict. McDermott notes, however, that very little of this research has focused on individuals. Most explanations for conflict find their origin in some aspect of the international environment or within particular state systems. Yet these institutions and organizations are comprised of individuals who, acting alone and in concert, often instigate or ameliorate situations of conflict and war. At this individual level of analysis, one of the most universal regularities across species and cultures lies in the fact that men engage in all kinds of physical conflict and fighting to a much greater degree than do women.

In exploring why this surprisingly neglected fact may be the case, McDermott and her collaborators undertook a series of experiments using simulated war games. As with many experimental research agendas, one set of findings inevitably leads to new questions and new studies designed to test and probe the lacuna left from the previous work. As attempts to distinguish the social and biological bases of aggression have continued, these experiments have become much more biological in nature than originally expected. As a result, unexpected and important ethical challenges have arisen as well. After briefly outlining the history of this work, the McDermott chapter concentrates on some of the specific ethical issues that arise when an investigator collects biological materials, particularly material that inherently includes identifying genetic material from the participants.

Kevin S. Reimer, Michael L. Spezio, Warren S. Brown, Gregory R. Peterson, James Van Slyke, and Kristen Renwick Monroe are part of a new generation of scholars working in moral psychology. In their chapter, Reimer et al. present a new approach to the study of moral courage, long the stuff of stories and myths. By drawing on recent work in moral psychology and decision neuroscience, Reimer and his colleagues examine both exemplary and ordinary manifestations of courage, investigating them via schemas, emotions, and appraisals, using a computational knowledge representation model known as latent semantic analysis (LSA) to distinguish narratives of people who demonstrated exemplary versus ordinary acts of courage during the Holocaust. Holocaust rescuers recognized for moral courage show a close cognitive association between actions and goals relative to soldiers who fought during World War II. The method anticipates studies of other types of courageous behavior and allows for increased external validation of laboratory studies informed by goal-oriented, recursive interactions between emotion and appraisal. It reflects the future in work on neuroscience and ethics and forges important ties across disciplines to explicate an age-old ethical problem: why do some people risk their lives to save others?

Jennifer L. Hochschild addresses a classic moral conundrum at the heart of liberal democratic politics: should a person choose what is best for him or herself or what best advances the interests of the group to which he or she belongs? Hochschild explores this conundrum in three related cases—skin color hierarchy among African Americans, multiracialism, and genomics. Each case offers possibilities for blurring, crossing, or even dissolving racial boundaries as they have been understood in the United States for most of the past century. Hochschild argues that any such change in a racial boundary might benefit the individual who makes it and might also diminish the strength or cohesiveness of that person's group, especially if he or she identifies as African American. Hochschild concludes by showing how and why the conundrum could occur in each of the three cases. She then identifies political situations and policy choices that can exacerbate, or soften, the potential dilemma of having to choose between individual or group benefits. Her work provides new insight into racial issues both traditionally and broadly conceptualized, and holds the potential to restructure discussions of both racial and group politics in the United States.

The next chapter applies social choice theory to the analysis of the ethical obligations of parents to children. Kenneth J. Arrow argues that in secular discourse, especially that associated with economic and, more generally, consequentialist thinking, all individuals, including children, have rights and obligations induced by the need to preserve the benefits of society. Despite this, the literature in this field places no special emphasis on the parent-child relation. Even religious discourse, though emphatic on the obligations of children to parents, places little emphasis on the reciprocal obligations. In fact, of course, a great deal of social activity, including economic activity, consists of parental support and maintenance of children. Moreover, the future adult life of children is profoundly affected by the family environment, so parental behavior becomes a serious externality for children. Indeed, the legal systems of most countries recognize these problems, if only in an ad hoc manner. Arrow examines the questions of obligation that are raised by a consideration of these issues, suggesting that there are real problems in the parent-child relation and in the legal system surrounding it. Arrow's chapter reveals the richness of social choice theory and economic analysis as tools to explicate ethical issues.

Thomas Schelling also demonstrates the power of economic analysis to yield insight into ethical concerns, in this instance the issue of global warming and climate change. Forsaking the ideological approaches of either the hard-line environmentalist or the purely profit-driven businessman, Schelling uses tools of decision theory to ask how we can realistically address environmental change. His analysis attempts to allow for the political realities of the market and political establishments, as well as for the needs of all human beings, both those scattered around the world and those not yet born, all of whom will be affected by our decisions. Schelling characterizes the climate change problem as a "bundle of uncertainties," among which are the amount by which the average global temperature may rise; the impact on particular regions, such as the high mountains; and how climate change may affect human life in the future, given our constantly evolving technologies. Despite the uncertainties, Schelling concludes that we can be sure that climate change will have the most deleterious impact on less-developed countries. Schelling predicts that successfully addressing climate change will require an unprecedented level of cooperation among the United States and other major nations, perhaps on the scale of the North Atlantic Treaty Organization.

Cheryl Koopman and Nicole Wernimont examine the psychological consequences of political violence in an international context. Their chapter highlights the ethical considerations that accompany increased awareness of this body of knowledge, which clearly and consistently links human suffering and psychological damage to politically-motivated violence. Koopman and Wernimont detail how various forms of political violence—whether genocide, terrorism, torture, kidnapping, or the situation of refugees—frequently result in psychological trauma manifesting itself in post traumatic shock disorder syndrome (PTSD), traumatic memories, dissociation, changes in world view, or increased sensitization to future stress. The authors argue that research in this field is now so overwhelming that psychological science has an imperative to raise the question of ethical responsibility in the face of political violence. Koopman and Wernimont call for practitioners of this kind of research to share their findings with the public and with policymakers, who must in turn accept their obligation to work to prevent such violence in the future.

Joseph F. C. DiMento and Gilbert Geis consider extraordinary rendition—sometimes labeled torture by proxy—and show how the legal judgments of the practice have reached varying conclusions. The authors first present four brief profiles of instances of extraordinary rendition and legal responses to them. Next, DiMento and Geis discuss international law on extraordinary rendition, including treaties such as the United Nations Convention against Torture and Other Cruel, Inhuman or Degrading Treatment or Punishment (CAT) and the relevant domestic American law, ranging from both statutory provisions dealing with rendition and torture and constitutional law regarding the authority of the Executive. In addition to their legal analysis, DiMento and Geis discuss both the morality of extraordinary rendition and its effectiveness in extracting valuable and reliable information. They emphasize that there is no evidence that the practice produces useful intelligence. Further, they argue that even if legal rendition were to produce reliable useful information, the means being used to that end are ethically intolerable. They make a strong case that while some might seek to justify the policy in jurisprudential terms, legal rendition is both legally highly suspect and morally reprehensible, and it is a procedure that brings discredit and shame on the United States. The authors conclude by laying out several routes aimed at ending the practice of extraordinary rendition.

Kristen Renwick Monroe and William Chiu turn their attention to what is fast becoming one of the major recognized problems facing the world: equality for women. They ask why, after some thirty years of governmental policies designed to advance the position of women, women in the United States still find such a gap between the wages and positions available to women versus men in the labor force. The usual answer provided involves the "pipeline" model, a view that argues that even with a level playing field provided by unprejudiced access to education and training, it simply takes time for women who enter educational and labor markets to advance to the top of their fields, or pass through the "pipeline." In fact, as Monroe and Chiu demonstrate using aggregate data, this is not occurring. More women indeed are entering the field, but few make it to the top; the pipeline is a leaky one and the people who leak out of the system tend to be female. Monroe and Chiu's analysis considers both aggregate US data and data focusing on academia itself. The authors conclude that academia reflects the pattern found in the broader workplace; they end with a discussion of how work that presents such evidence is received by the powers that be in the Academy.

Each of these chapters highlights the importance of science—broadly conceptualized—in the elucidation of ethical issues and suggests both how a scientific approach can produce new insight on eternal issues and how there are political concerns that are woven into our ethical analyses. We hope it will stimulate future work in this direction and can be used in ethics courses, alongside works following more traditional approaches.

Chapter 1

Evolutionary Biology and the Origin of Morality

Francisco J. Ayala, Kristen Renwick Monroe, and Nicholas Lampros

Conversation

Kristen Monroe (KM): In your opinion, what is it that sets humans apart from other animals?

Francisco Ayala (FA): We are aware that we exist as individuals and we are aware that we will die. Animals are not self-aware. If we are self-aware, that is, if we know we exist as individuals, we know we'll die, because all other members of our species die. Self-awareness implies death-awareness. Any creature that has death-awareness will treat the dead of its own species with respect and will take care of the dead because he or she knows that he or she is going to die and wants to be treated the same way. The only species that practices ceremonial burying of the dead is human beings. That indicates that only humans have death-awareness, which in turn implies that only humans have self-awareness. This connection is very interesting, because I think one way that we may learn when morality started and when self-awareness and language started is by discovering when ceremonial burial of the dead started in the hominids, in our evolution. We know that Neanderthals and early *Homo sapiens* were practicing ceremonial burial of the dead. As to earlier hominids, it is not clear, not yet—eventually more precise information might be discovered.

KM: What about stories of dolphins who recognize themselves in mirrors? Or chimpanzees?

FA: They do to a haphazard and limited extent. All animal lovers think I am being too harsh on other animals. They say dogs and other animals show intelligence. Other animals have intelligence, but not to the extent that we do; it is not abstract and rational intelligence. Concerning the work on dolphins—well, you are referring to some experiments that were first done with chimps and very soon thereafter with crows, but also with dolphins. First of all, they train a chimp to be in front of a mirror. Chimps at first don't seem to be aware that what they are seeing is themselves. When they are sleeping, the trainers put a spot on their foreheads. When they wake up and look in the mirror, it seems that, in enough cases to be intriguing, the chimp, after playing for a while and touching the spot in the mirror, sometimes realizes that the spot is on him, not on some other guy. Well, crows and dolphins apparently do fairly well with this kind of self-recognition test. Apparently crows fairly soon recognize that they are looking at themselves. And dolphins are one of the most intelligent animals; they are mammals. They evolved great intelligence. So there is the beginning of self-recognition, some degree of recognizing that something is associated with dolphins and with chimps and with other animals. But death awareness is a necessary consequence of self-awareness, and death awareness is evinced by ceremonial burial, which only humans practice.

KM: You're saying that the way to prove self-awareness is to look at the manifestation of the knowledge that we are going to die, and that this awareness is what distinguishes human beings from other higher species?

FA: Yes. Lack of death-awareness and ceremonial burial of the dead is the way to prove that other animals don't have self-awareness, because if you have self-awareness, you will have death-awareness. Of course, being self-aware implies much more than being death-aware.

KM: In this area, what are some of the big controversies and big questions that you would like to see answered?

FA: You always ask easy questions! Well, I'm disturbed by the conclusion drawn by sociobiologists and some philosophers, like Michael Ruse, who assert that free will is fantasy. Michael Ruse says that morality exists as a trick of our genes, which make us believe there is morality so that we behave according to what is good for our genes, even if we don't know it. I find this very shocking. So I would like to see it recognized that free will is real, that morality is real, and that the norms of morality are not determined by our genes. The same way that our minds evolved, we should evolve our moral principles, rather than sticking always to the same values. I would like to find out when morality, self-awareness, death-awareness, and language emerged in human evolution. That interests me. I have most recently published a book on human evolution.[1] Most books on human evolution don't give a lot of space to these topics, but this is one of the books that does. The "uniqueness of being human" is one chapter, which is sixty pages long. This book was published in November or December 2007, and already in March 2008 they said they needed a second printing; they were selling it so fast. So people are interested in these topics. They're important.

KM: Let's go back to questions of free will. You mentioned Michael Ruse, and much of his explanation seems rather tautological to me. It's hard to get evidence to bear on it. I just wondered what kind of evidence you feel you have regarding free will?

FA: Well, free will is a difficult problem. My first, simple argument is based on the experience we have that free will is genuine, not just a mirage. I am speaking with you by my free choice. But it is a major topic that deserves exploration in depth.

KM: The second set of questions you would like to see work on is when morality, self-awareness, burials, and language evolved. Are those really questions that anthropologists and archaeologists would be interested in?

FA: That's the point I was making. I want the people who take courses in anthropology to raise those questions. Some of them are interested in these questions, but we need to bring them more to the forefront. To what extent did *Homo erectus* or other hominid species practice burial of the dead? What kind of artistic representations did they make? These things are related. There is a very gradualistic view of the evolution of human intelligence and behavior that is advanced by someone called Terrence Deacon.[2] He is one of the most articulate proponents of the gradual evolution of morality and other distinctively human features. Like going from here to San Francisco, you go one step at a time. The prevailing view among some anthropologists is that there was a really quick quantum jump. That is tantamount to saying that the evolution of language, morality, art, and full rationality was fairly quick. I am in-between. I think it was probably more gradual than the second group believes but not as gradual as people like Terry Deacon and others think. There had to be some sort of threshold. You need a certain degree of intelligence before language, morality, and representation are present. Somehow around two million years ago, our ancestors started to produce tools. In order to make a tool, you have to have a fair degree of intelligence: you have to anticipate the future; you have to anticipate what uses you will make of the tool. Forming mental images of future events, making imaginative representations of realities that are not present, is a uniquely human trait. Many animal lovers think otherwise. I think we humans are the only ones who can do that. *Homo habilis,* our ancestors of two million years ago, started to make tools, which became very beneficial—those who had tools survived and reproduced better, and therefore those who were more intelligent and were more likely to make tools were having more children, and thus the genes for intelligence were increasing in frequency.

KM: Was morality a by-product?

FA: Morality was a by-product much later. It's not clear to me that in the early stages of tool making there was broad anticipation of the future, which is implied in moral behavior—anticipation of the consequences of our actions. Again, you have to reach a threshold in my view before you can really say that somebody is behaving morally. This is all related to self-awareness. In order to make a tool, you have to anticipate the future, and you have to have a sense of yourself. But the capacity to anticipate the future as well as self-awareness developed somewhat gradually, not suddenly when the first tools were made.

KM: I know you've studied Aquinas and Aristotle. Aristotle said man is a social being. That's very much what I hear you saying.

FA: Yes. I am very interested in Aristotle and Aquinas, who was influenced by Aristotle.

KM: Was your time in the church relevant to your later work?

FA: I studied theology for five years, so I learned Aquinas, and also I studied philosophy in Spain—you know I studied philosophy for two or three years. Spanish philosophers in the fifties were very scholastic and in the Greek tradition. They were teaching Aristotle and Plato, and the modern version of Aristotle and Plato. Then in theology, Aquinas was king. I published my first book, my dissertation, on human evolution. My thesis of theology was written in Latin. It was on the teachings of the Sacred Scriptures about human origins. I don't know that anyone has read the book. But my philosophy studies left a lasting influence on me.

KM: You wrote the book in Latin?

FA: Yes.

KM: You were born in Spain?

FA: Yes. I thought I was born in heaven but maybe it was in Spain. Madrid.

KM: You went to what you call high school in Madrid?

FA: Yes.

KM: What happened after that? You went to seminary?

FA: I went to a Catholic high school. Priests were the teachers. We had some lay teachers, but mostly they were priests. The private schools in Spain were Catholic; Spain was officially Catholic. I finished high school by 1950. I went then to study physics at the University of Madrid. I got an undergrad degree in physics, and then before I was finished with physics I started to study philosophy. When I finished my studies in philosophy, I started to study theology, and that was when I entered the church formally, as it were, as a Dominican monk.

KM: So this was around 1960?

FA: I finished my theology studies in 1960. By that time I had become very interested in evolution and genetics. I had decided to come to the United States to study genetics and evolution. In 1961, I came to Columbia University to work with a very distinguished evolutionist, Theodosius Dobzhansky, and he was the one who persuaded me to stay in this country. When I first came, I had no intention whatsoever of staying here.

KM: You are the only person I know who is in three different schools.

FA: I have the title of University Professor. With the title of University Professor, I can teach in any campus throughout the UC system. Initially I was a Distinguished Professor. Now, I like to say, I became "undistinguished" because I became the Donald Bren Professor of Biological Sciences, though I think nominally I still have the right to the title of "distinguished." So, now I am the "University Professor" here. There can only be one University Professor at a campus, I believe. But also, as you say, at UCI I am professor in three schools: Biological Sciences, Social Sciences, and Humanities.

KM: What would you like your legacy to be?

FA: I don't think about having a legacy. I am aware of my existence and I have death-awareness. There is always a little bit of hope that death might bypass me, I like to say playfully, since I am an unrepentant optimist. But seriously, this may surprise you, but I don't aspire to have any particularly great or lasting legacy. I think leading a full life with my family, with my friends, with my students, and sharing whatever I can share is enough to fulfill my life. I don't think about the future after my death; I don't worry whether my books or papers will be cited fifty years from now, I really don't.

KM: Are science and religion incompatible? I know you've gotten involved in that argument.

FA: This debate seems very unfortunate to me. Most important perhaps is that this juxtaposition makes very religious people unhappy and even unwilling to acquire a proper scientific education. My argument is that evolution is actually more compatible with religion than the so-called theory of Intelligent Design. I am very much in favor of scientific education. I am very much in favor of religion; I think religion in general is something that brings happiness and satisfaction to billions of people in the world. Why would we want to have religion eradicated from the world, as some writers have proposed? Why would you want to take religion away from people? People suffer from hunger, from disease. Religion is the one thing that brings hope and satisfaction to millions of people. I have nothing against religion. Quite the contrary. Some authors refer to bad deeds that were done in history in the name of religion. Science has done bad things too, but this is no reason to be against science.

What I am trying to say in some of my books is something more, namely about this theory of Intelligent Design—well, if God has designed human beings and the rest of the living world, then God has a lot to account for. He has designed a jaw that is not big enough for the teeth, so we have to remove the wisdom teeth. You can look at almost anything in human anatomy and find very defective design. Take our eye. Because of evolution, the optical nerve forms inside the cavity and now the nerve has to cross the retina to go to the brain, so we have a blind spot. The octopus and the squid have an eye as complex as ours, but it does not have a blind spot. So God loves the squid and octopus more than us? Is that what the Intelligent Design proponents are implying? The female birth canal is not big enough for the head of the baby—it has to do with the enlargement of our head. So, many children die as a consequence. Twenty percent of all human pregnancies end in spontaneous abortion, miscarriages that occur in the first two months. That's twenty million *children* per year in the world for religious people who believe that human life begins at conception. There are twenty million dead babies every year because the human reproductive system is poorly designed. Are you going to blame God for that? I say Intelligent Design is blasphemy. But the human reproductive system is the outcome of evolution, the outcome of natural processes. There's no morality there. But those who claim that humans, like other living beings, have been specifically designed by God, are blaming God for the bad design and for the twenty million spontaneous abortions per year in the world.

KM: Are you yourself religious?

FA: I don't answer that question. I never answer that. No matter what I say, one side or the other will attribute my ideas to my religious faith or my lack of it. I think my ideas are independent of whether or not I am a person of faith.

NOTES

1. Cela-Conde and Ayala 2007.
2. Deacon 1998.

REFERENCES

Cela-Conde, Camilo J., and Francisco J. Ayala. 2007. *Human Evolution: Trails from the Past.* Oxford: Oxford University Press.

Deacon, Terrence. 1998. *The Symbolic Species: The Co-Evolution of Language and the Brain.* New York: W. W. Norton & Co.

The Design of Organisms

From Paley to Darwin

Francisco J. Ayala

Investigation

I vividly remember the day in 1971 when it was announced in New York that the Metropolitan Museum of Art had acquired at auction in Christie's London headquarters the painting *Juan de Pareja* by Diego Velázquez. The Metropolitan had paid the staggering sum of $5,544,000, more than had ever before been paid for a painting, no matter how illustrious the artist or distinguished the work. Thomas Hoving, the Metropolitan's director at the time, later referred to *Juan de Pareja* as "the most important painting in world history" (1993, 254). This is hyperbole, but there can be little doubt that this painting is one of the finest portraits crafted by the great Spanish master. In the summer of 1649, Velázquez had brought with him from Madrid to Rome his Moorish servant and painted him while getting ready to portray Pope Innocent X. The *Juan de Pareja* is distinctively Velázquez's in the long but firm brushstrokes, the clarity of execution, and in the color palette, with the body executed in browns textured with brilliant reds and blacks and set dramatically aside from the face by a large "white" shirt collar extending from neck to shoulder. Like other Velázquez portraits of this period, form is created by color. Velázquez is said to have made a point of sending Pareja to visit his friends while carrying the portrait so as to astonish them with its vivid likeness. In the spring of 1650, the painting was exhibited to great critical acclaim in Rome's Pantheon, and this success may have been a reason for Velázquez's election to the Roman Academy later that year.

In distant northern India a few years later, an unknown Muslim craftsman forged the *Dagger of Aurangzeb,* one of the most exquisite treasures in the great collection of Indian decorative arts held by the Los Angeles County Museum of Art. The hilt, in the form of a horse's head and neck, is crafted from green jade highlighted in dark orange. The blade, made of damascened steel, is shaped in the curvilinear "khanjar" style and exhibits floral ornaments and an inscription inlaid in gold with the date 1660/61. The Taj Mahal had just been built at Agra by the Mogul emperor Shah Jahan as a mausoleum to honor his dead wife, Mumtaz Mahal. The *Dagger of Aurangzeb* was designed as a weapon, but also as a decorative object; its author was a refined artist, not just a craftsman.

It would be ridiculously misplaced to suggest that the *Dagger of Aurangzeb* was a product of chance rather than design. It would be equally ridiculous to suggest that Velázquez executed *Juan de Pareja* without any preconceived plan or purpose. The exquisite design of organisms and their marvelous adaptations were similarly explained, up to the mid-nineteenth century, as an outcome of the plan and purpose of a Creator. God had created the birds and bees, the fish and corals, the trees in the forest, and best of all, man. God had given us eyes so that we might see, and He had provided fish with gills to breathe in water. Scientists, philosophers, and theologians argued that the complex and exquisite design of organisms manifests indeed the existence of an all-wise Creator. Wherever there is design, there is a designer; the existence of a watch evinces the existence of a watchmaker.

The English theologian William Paley in his *Natural Theology* (1802), for example, elaborated the argument-from-design as a forceful demonstration of the existence of the Creator. The functional design of the human eye, argued Paley, provides conclusive evidence of an all-wise Creator. It would be absurd to suppose, he wrote, that by mere chance the human eye "should have consisted, first, of a series of transparent lenses ... secondly of a black cloth or canvas spread out behind these lenses so as to receive the image formed by pencils of light transmitted through them, and placed at the precise geometrical distance at which, and at which alone, a distinct image could be formed ... thirdly of a large nerve communicating between this membrane and the brain" (65). The *Bridgewater Treatises*, published between 1833 and 1840, were written by eminent scientists and philosophers to set forth "the Power, Wisdom, and Goodness of God as manifested in the Creation." Thus, the distinguished anatomist and surgeon Sir Charles Bell (1833) described the fanciful structure and mechanisms of the human hand as incontrovertible evidence that the hand had been designed by the same omniscient Power that had created the world.

COPERNICUS AND DARWIN

There is a version of the history of ideas that sees a parallel between the Copernican and the Darwinian Revolutions. In this view, the Copernican Revolution consisted of displacing the Earth from its previously accepted locus as the center of the universe, and moving it to a subordinate place as just one more planet revolving around the sun. In congruous manner, the Darwinian Revolution is viewed as consisting of the displacement of humans from their exalted position as the center of life on Earth, with all other species created for the service of humankind. According to this version of intellectual history, Copernicus had accomplished his revolution with the heliocentric theory of the solar system. Darwin's achievement emerged from his theory of organic evolution. Most important about these two intellectual revolutions, however, is that they ushered in the beginning of science, in the modern sense of the word. These two revolutions may jointly be seen as the one Scientific Revolution with two stages, the Copernican and the Darwinian.

The Copernican Revolution was launched with the publication in 1543—the year of Nicolaus Copernicus's death—of his *De revolutionibus orbium celestium* (*On the Revolutions of the Celestial Spheres*) and bloomed with the publication in 1687 of Isaac Newton's *Philosophiae naturalis principia mathematica* (*The Mathematical Principles of Natural Philosophy*). The discoveries by Copernicus, Kepler, Galileo, Newton, and others in the sixteenth and seventeenth centuries had gradually ushered in a conception of the universe as matter in motion governed by natural laws. It was shown that Earth is not the center of the universe, but a small planet rotating around an average star, that the universe is immense in space and in time, and that the motions of the planets around the sun can be explained by the same simple laws that account for the motion of physical objects on our planet.

These and other discoveries greatly expanded human knowledge. The conceptual revolution they brought about was more fundamental yet: a commitment to the postulate that the universe obeys immanent laws that account for natural phenomena. The workings of the universe were brought into the realm of science: explanation through natural laws. All physical phenomena could be accounted for as long as the causes were adequately known.

The advances of physical science brought about by the Copernican Revolution had driven mankind's conception of the universe to a split-personality state of affairs, which persisted well into the mid-nineteenth century. Scientific explanations, derived from natural laws, dominated the world of nonliving matter, on the Earth as well as in the heavens. Supernatural explanations, such as Paley's explanation of design, which depended on the unfathomable deeds of the Creator, accounted for the origin and configuration of living creatures—the most diversified, complex, and interesting realities of the world.

It was Darwin's genius to resolve this conceptual schizophrenia. Darwin completed the Copernican Revolution by drawing out for biology the notion of nature as a lawful system of matter in motion that human reason can explain without recourse to supernatural agencies. The conundrum faced by Darwin can hardly be overestimated. The strength of the argument from design to demonstrate the role of the Creator had been forcefully set forth by William Paley in his *Natural Theology* (1802). Wherever there is function or design, we look for its author. A knife is made for cutting and a clock is made to tell time; their functional designs have been contrived by a knifemaker and a watchmaker. The exquisite design of

Velázquez's *Juan de Pareja* proclaims that it was created by a gifted artist following a preconceived plan. Nor is the *Dagger of Aurangzeb* an accident of nature or a random assemblage of materials, but the product of design. Similarly, the structures, organs, and behaviors of living beings are directly organized to serve certain purposes or functions. The functional design of organisms and their features would therefore seem to argue for the existence of a Designer. It was Darwin's greatest accomplishment to show that the complex organization and functionality of living beings can be explained as the result of a natural process—natural selection—without any need to resort to a Creator or other external agent. The origin and adaptation of organisms in their profusion and wondrous variations were thus brought into the realm of science.

Darwin accepted that organisms are "designed" for certain purposes; that is, they are functionally organized. Organisms are adapted to certain ways of life and their parts are adapted to perform certain functions. Fish are adapted to live in water; kidneys are designed to regulate the composition of blood; the human hand is made for grasping. But Darwin went on to provide a natural explanation of the design. The seemingly purposeful aspects of living beings could now be explained, like the phenomena of the inanimate world, by the methods of science, as the result of natural laws manifested in natural processes.

DESIGN WITHOUT DESIGNER

It is my contention that Darwin's most revolutionary achievement is that he extended the Copernican Revolution to the world of living things, much more so than his demonstration of the evolution of organisms. Henceforward, the origin and design of organisms could be explained, like the phenomena of the inanimate world, as the result of natural laws manifested in natural processes. Darwin's theory encountered opposition in some religious circles, not so much because he proposed the evolutionary origin of living things (which had been proposed before and accepted even by Christian theologians), but because the causal mechanism, natural selection, excluded God as the explanation for the design of organisms. The Roman Catholic Church's opposition to Galileo in the seventeenth century had been similarly motivated, not only by the apparent contradiction between the heliocentric theory and a literal interpretation of the Bible, but also by the unseemly attempt to comprehend the workings of the Universe, the "mind of God." The configuration of the Universe was no longer perceived as the result of God's Design, but simply the outcome of immanent, blind processes.

There were, however, many theologians, philosophers, and scientists who saw no contradiction then, nor see it now, between the evolution of species and Christian faith. Some see evolution as the "method of divine intelligence," in the words of the nineteenth century theologian A. H. Strong (1886, 237). Others, like the American contemporary of Darwin, Henry Ward Beecher (1818–1887), made evolution the cornerstone of their theology.

These two traditions have persisted to the present. Pope John Paul II has stated (October 1996) that "the theory of evolution is more than a hypothesis. It is ... accepted by researchers, following a series of discoveries in various fields of knowledge." And there is a plethora of "process" theologians, who perceive evolutionary dynamics as a pervasive element of a Christian view of the world.

The evolution of organisms was commonly accepted by naturalists in the middle decades of the nineteenth century. The distribution of exotic species in South America, in the Galápagos Islands, and elsewhere, as well as the discovery of fossil remains of long-extinguished animals, confirmed the reality of evolution in Darwin's mind. The intellectual challenge was to explain the origin of distinct species of organisms and how new ones became adapted to their environments, that "mystery of mysteries," as it had been labeled by Darwin's older contemporary, the prominent scientist and philosopher Sir John Herschel (Cannon 1961, 301).

Early in his notebooks of 1837 to 1839, which he had started shortly after returning from a five-year trip around the world in the HMS *Beagle,* Darwin registered his discovery of natural selection and thereafter, over the years, he repeatedly referred to it as "my theory." From the late 1830s until his death in 1882, Darwin's life would be dedicated to substantiating natural selection and its companion postulates, mainly the pervasiveness of hereditary variation and the enormous fertility of organisms, which much surpassed the capacity of available resources. Natural selection became for Darwin "a theory by which to work" (1958, 120). He relentlessly pursued observations and performed experiments in order to test the theory and resolve presumptive objections.

Alfred Russel Wallace (1823–1913) is famously given credit for discovering, independently of Darwin, natural selection as the process accounting for the evolution of species. On June 18, 1858, Darwin wrote to Charles Lyell that he had received by mail a short essay from Wallace such that "if Wallace had my [manuscript] sketch written in [1844] he could not have made a better abstract" (1991, 107). Darwin was thunderstruck.

Wallace's independent discovery of natural selection is remarkable. Wallace, however, was not interested in explaining design, but rather in accounting for the evolution of species, which he saw as a sustained and progressive process, as indicated in his paper's title: "On the Tendency of Varieties to Depart Indefinitely from the Original Type." Wallace thought that evolution proceeds indefinitely and is progressive. Darwin, on the contrary, did not accept that evolution would necessarily represent progress or advancement, nor did he believe that evolution would always result in morphological change over time; rather, he knew of the existence of "living fossils," organisms that had remained unchanged for millions of years. For example, "some of the most ancient Silurian animals, as the Nautilus, Lingula, etc., do not differ much from living species" (1859, 336). (We now know that the Silurian geological period lasted from 444 to 416 million years ago.)

In 1858, Darwin was at work on a multivolume treatise, intended to be titled "On Natural Selection." Wallace's paper stimulated Darwin to write *The Origin,* which would be published the following year. Darwin intended this as an abbreviated version of the much longer book he had planned to write. Darwin's focus, in *The Origin* as elsewhere, was the explanation of design, with evolution playing the subsidiary role of supporting evidence.

NATURAL SELECTION

The central argument of the theory of natural selection was summarized by Darwin in *The Origin of Species* as follows:

> As more individuals are produced than can possibly survive, there must in every case be a struggle for existence, either one individual with another of the same species, or with the individuals of distinct species, or with the physical conditions of life.... Can it, then, be thought improbable, seeing that variations useful to man have undoubtedly occurred, that other variations useful in some way to each being in the great and complex battle of life, should sometimes occur in the course of thousands of generations? If such do occur, can we doubt (remembering that more individuals are born than can possibly survive) that individuals having any advantage, however slight, over others, would have the best chance of surviving and of procreating their kind? On the other hand, we may feel sure that any variation in the least degree injurious would be rigidly destroyed. This preservation of favorable variation and the rejection of injurious variations, I call Natural Selection. (1967, 63; 80–81)

Darwin's argument addresses the problem of explaining the design of organisms. Darwin argued that adaptive variations ("variations useful in some way to each being") occasionally appear and that these are likely to increase the reproductive chances of their carriers. Over the generations, favorable variations will be preserved; injurious ones will be eliminated. In one place, Darwin added: "I can see no limit to this power [natural selection] in slowly and beautifully adapting each form to the most complex relations of life" (1859, 469). Natural selection was proposed by Darwin primarily to account for the adaptive organization, or "design," of living beings; it is a process that promotes or maintains adaptation, which calls for precisely functioning organisms. Evolutionary change through time and the multiplication of species are not directly promoted by natural selection, although they typically ensue as by-products of natural selection fostering adaptation. The point merits emphasis: natural selection directly promotes the adaptation of organisms to their environments, not evolution, although evolutionary change often follows as a consequence, due in part to the diversity of environments in which organisms live and to the continuously changing conditions of the environments.

Darwin formulated natural selection primarily as differential survival. The modern understanding of the principle of natural selection is formulated in genetic and statistical terms as differential reproduction. Natural selection simply implies that some genes and genetic combinations are transmitted to the following generations more frequently than their alternates. Favored genes will become more common in every subsequent generation and their alternates less common. Natural selection is a statistical bias in the relative rate of reproduction of alternative genes.

CREATION: GOD'S, ART'S, NATURAL SELECTION'S

Natural selection has been compared to a sieve that retains the rarely arising useful genes and lets go of the more frequently arising harmful mutants. Natural selection acts in that way, but it is much more than a purely negative process, for it is able to generate novelty by increasing the probability of otherwise extremely improbable genetic combinations. Natural selection is thus creative in a way. It does not "create" the entities upon which it operates, but it produces adaptive genetic combinations that would not have existed otherwise.

The creative role of natural selection must not be understood in the sense of the "absolute" creation that traditional Christian theology predicates of the Divine act by which the universe was brought into being ex nihilo. Rather, natural selection may be compared to a painter who creates a picture by mixing and distributing pigments in various ways over the canvas. The canvas and the pigments are not created by the artist but the painting is. It is conceivable that a random combination of pigments might result in the orderly whole which is a work of art. But the probability of Velázquez's *Juan de Pareja* resulting from a random combination of pigments, or Disneyland's Magic Mountain resulting from a random association of bricks, concrete, and other materials, is infinitely small. In the same way, the combination of genes that carries the hereditary information responsible for the formation of the human eye could have never been produced by a random process. Not even if we allow for the three billion-plus years during which life has existed on Earth. The complicated anatomy of the eye, like the exact functioning of the kidney, is the result of a nonrandom process—natural selection.

Critics of evolution have alleged as evidence against Darwin's theory of evolution examples showing that random processes cannot yield meaningful, organized outcomes, such as the human eye or the complex biochemistry of cells. It is pointed out that a series of monkeys randomly striking letters on a typewriter would never write *The Origin of Species,* even if we allow for millions of years and many generations of monkeys pounding at typewriters.

This criticism would be valid if evolution depended only on random processes. But natural selection is a nonrandom, or "directional," process that promotes adaptation by selecting combinations that "make sense," that is, that are useful to the organisms. The analogy of the monkeys would be more appropriate if a process existed by which, first, meaningful words would be chosen every time they appear on the typewriter, and then we would also have typewriters with previously selected words rather than just letters in the keys, and again there would be a process to select meaningful sentences every time they appear in this second typewriter. If every time words such as "the," "origin," "species," and so on appeared in the first kind of typewriter, they each became a key in the second kind of typewriter, meaningful sentences would occasionally be produced in this second typewriter. If such sentences became incorporated into keys of a third type of typewriter, in which meaningful paragraphs were selected whenever they appear, it is clear that pages and even chapters "making sense" would eventually be produced.

We need not carry the analogy too far, since no analogy is fully satisfactory, but the point I wish to make should be clear. Evolution is not the outcome of purely random processes, but rather there is a "selecting" process, which picks up adaptive combinations because these reproduce more effectively and thus become preserved and multiplied. These adaptive combinations constitute, in turn, new levels of organization upon which the mutation (random) and selection (nonrandom or directional) processes again operate. Mutations are genetic changes that occur whether or not they are useful to the organisms and, in this sense, they are random events. But natural selection promotes the useful and eliminates the harmful.

THE SUCCESS OF THE IMPROBABLE

The manner in which natural selection can generate novelty in the form of accumulated hereditary information may be illustrated by the following example. Some strains of the colon bacterium *Escherichia coli,* in order to be able to reproduce in a culture medium, require that a certain substance, the amino acid histidine, be provided in the medium. When a few such bacteria are added to a small laboratory tube containing ten cubic centimeters of liquid-culture medium containing histidine, the bacteria multiply rapidly and produce about twenty billion bacteria in just one day or two, depending on the temperature and other conditions. Spontaneous (random) mutations to streptomycin resistance occur in normal (i.e., sensitive) bacteria at rates of the order of one in one hundred million (1×10^{-8}) cells. In the bacterial culture, we expect about two hundred bacteria to be resistant to streptomycin due to spontaneous mutation. If a proper concentration of the antibiotic is added to the culture, only the resistant cells survive. The two hundred surviving bacteria will start reproducing, however, and, allowing a day or two for the necessary number of cell divisions, many billion bacteria are produced, all resistant to streptomycin.

Among cells requiring histidine as a growth factor, spontaneous mutants able to reproduce in the absence of histidine arise at random, at rates of about four in one hundred million (4×10^{-8}) bacteria. The streptomycin-resistant cells may now be transferred to a culture with streptomycin but with no histidine. Most of them will not be able to reproduce, but about several hundred will start reproducing until the available medium is saturated.

Natural selection has produced, in two steps, bacterial cells resistant to streptomycin and not requiring histidine for growth. The probability of the two mutational events happening in the same bacterium is about four in ten million billion ($1 \times 10^{-8} \times 4 \times 10^{-8} = 4 \times 10^{-16}$) cells. An event of such low probability is unlikely to occur even in a large laboratory culture of bacterial cells. With natural selection, cells having both properties are the common result. All the billion bacteria now living in the small tube have adapted to the particular conditions prevailing in the laboratory culture: presence of the antibiotic streptomycin and absence of the amino acid histidine.

As illustrated by the bacterial example, natural selection produces combinations of genes that would otherwise be highly improbable because natural selection proceeds stepwise. The human eye did not appear suddenly in all its present perfection. Its formation requires the appropriate integration of many genes, and thus the eye could not have resulted from random processes alone. Our ancestors had, for more than half a billion years, some kinds of organs sensitive to light. Perception of light, and later vision, were important for these organisms' survival and reproductive success. Accordingly, natural selection favored genes that increased the functional efficiency of the eye. Such genes gradually accumulated, until eventually the highly complex and efficient human eye emerged. Natural selection can account for the rise, multiplication, and combination of genes that are beneficial to the organisms in the environments where they live. Sunlight is a ubiquitous feature of the environment. Not surprisingly, more than three dozen different kinds of eyes have independently evolved in all sorts of animals. We know that they have evolved independently because they are so different in organization and mode of operation, although they all share perception of light, as well as some ancestral genes. No overall grand design here but, rather, opportunistic evolution. Look at an octopus's eye or at the compound eye of a fly, and you will notice how dramatically different they are from the eyes of humans and other vertebrates.

PURPOSE WITHOUT FORESIGHT

There is an important respect in which an artist makes a poor analogy of natural selection. A painter usually has a preconception of what he wants to paint and will consciously modify the painting so that it represents what he wants. Natural selection has no foresight, nor does it operate according to some preconceived plan. Rather it is a purely natural process resulting from the interacting properties of physicochemical and biological entities. Natural selection is simply a consequence of the differential multiplication of living beings. It gives some appearance of purposefulness because it is conditioned by the environment: which organisms reproduce more effectively depends on what variations they possess that are useful in the environment where the organisms live. But natural selection does not anticipate the environments of the future; drastic environmental changes may be insuperable to organisms that were previously thriving, and these may become extinct. More than 99 percent of all species that ever lived became extinct, leaving no descendant organisms.

The team of typing monkeys is also a bad analogy of evolution by natural selection because it assumes that there is "somebody" who selects letter combinations and word combinations that make sense. In evolution there is no one selecting adaptive combinations. These select themselves because they multiply more effectively than less adaptive ones.

There is a sense in which the analogy of the typing monkeys is better than the analogy of the artist, at least if we assume that no particular statement was to be obtained from the monkeys' typing endeavors but just any statements making sense. Natural selection does not strive to produce predetermined kinds of organisms, but only organisms that are adapted to their present environments. Which characteristics will be selected depends on which variations happen to be present at a given time in a given place. This in turn depends on the random process of mutation, as well as on the previous history of the organisms (i.e., on the genetic make-up they have as a consequence of their previous evolution). Natural selection is an "opportunistic" process. The variables determining in what direction it will go are the environment, the pre-existing constitution of the organisms, and the randomly arising mutations.

CHANCE AND OPPORTUNISM

Adaptation to a given environment may occur in a variety of different ways. An example may be taken from the adaptations of plant life to desert climate. The fundamental adaptation is to the condition of dryness, which involves the danger of desiccation. During a major part of the year, sometimes for several years in succession, there is no rain. Plants have accomplished the urgent necessity of saving water in different ways. Cacti have transformed their leaves into spines, having made their stems into barrels containing a reserve of water; photosynthesis is performed in the surface of the stem instead of in the leaves. Other plants have no leaves during the dry season, but after it rains they burst into leaves and flowers and produce seeds. Ephemeral plants germinate from seeds, grow, flower, and produce seeds—all within the space of the few weeks while rainwater is available; the rest of the year the seeds lie quiescent in the soil.

The opportunistic character of natural selection is also well evidenced by the phenomenon of adaptive radiation. The evolution of drosophila fruit flies in Hawaii is a simple example. There are about fifteen hundred drosophila species in the world. Approximately five hundred of them have evolved in the Hawaiian archipelago, although this is a small area, about one twenty-fifth the size of California. Moreover, the morphological, ecological, and behavioral diversity of Hawaiian drosophila exceeds that of drosophila in the rest of the world.

Why should such "explosive" evolution have occurred in Hawaii? The overabundance of drosophila flies there contrasts with the absence of many other insects. The ancestors of Hawaiian drosophila reached the archipelago before other groups of insects did, and thus they found a multitude of unexploited opportunities for living. They responded by a rapid adaptive radiation; although they are all probably derived from a single colonizing species, they adapted to the diversity of opportunities available in diverse places or at different times by developing appropriate adaptations, which range broadly from one to another species.

The process of natural selection can explain the adaptive organization of organisms, as well as their diversity and evolution, as a consequence of their adaptation to the multifarious and ever-changing conditions of life. The fossil record shows that life has evolved in a haphazard fashion. The radiations, expansions, relays of one form by another, occasional but irregular trends, and the ever-present extinctions are best explained by natural selection of organisms subject to the vagaries of genetic mutation and environmental challenge. The scientific account of these events does not necessitate recourse to a preordained plan, whether imprinted from without by an omniscient and all-powerful designer or resulting from some immanent force driving the process toward definite outcomes. Biological evolution differs from a painting or an artifact in that it is not the outcome of a design preconceived by an artist or artisan.

Natural selection accounts, though, for the "design" of organisms because adaptive variations tend to increase the probability of survival and reproduction of their carriers at the expense of maladaptive, or less adaptive, variations. The arguments of Paley and the authors of the *Bridgewater Treatises* against the incredible improbability of chance accounts of the origin of organisms are well taken, as far as they go. But neither these scholars, nor any other authors before Darwin, were able to discern that there is a natural process (namely, natural selection) that is not random but rather is oriented and able to generate order or "create." The traits that organisms acquire in their evolutionary histories are not fortuitous but determined by their functional utility to the organisms. It is unfortunate that some "creationists" repeat the arguments of yesteryear and proclaim that the designs of organisms are "intelligent" design, the products of special acts of creation. There is no more reason for that than there is for explaining the configuration of the continents, or the planets and the stars, as products of special acts of creation. Moreover, the arguments for intelligent design imply unacceptable religious consequences, as I shall argue.

THE CREATIVE DUET: CHANCE AND NECESSITY

Chance is, nevertheless, an integral part of the evolutionary process. The mutations that yield the gene variations available to natural selection arise at random, independently of whether they are beneficial or harmful to their carriers. But this random process (as well as others that come into play in the great theatre of life) is counteracted by natural selection, which preserves and multiplies what is useful and eliminates the harmful. Without mutation, evolution could not happen because there would be no variations that could be differentially conveyed from one to another generation. But without natural selection, the mutation process would yield disorganization and extinction, because most mutations are disadvantageous. Mutation and selection have jointly driven the marvelous process that, starting from microscopic organisms, has produced orchids, birds, and humans.

The theory of evolution manifests chance and necessity jointly intertwined in the stuff of life, randomness and determinism interlocked in a natural process that has spurted the most complex, diverse, and beautiful entities in the universe: the organisms that populate the Earth, including humans, who think and love, who are endowed with free will and creative powers, and who are able to analyze the process of evolution that brought them into existence. This is Darwin's fundamental discovery, that there is a process that is creative though not conscious. And this is the conceptual revolution that Darwin completed: that everything in nature, including the origin and design of living organisms, can be accounted for as the result of natural processes governed by natural laws. This is nothing if not a fundamental vision that has forever changed how mankind perceives itself and its place in the universe.

Paley's claim that the design of living beings evinces the existence of a Designer was shown to be erroneous by Darwin's discovery of the process of natural selection, just as the pre-Copernican explanation for the motions of celestial bodies (and the medieval argument for the existence of God based on the unmoved mover) was shown to be erroneous by the discoveries of Copernicus, Kepler, Galileo, and Newton.

The Copernican and Darwinian Revolutions have jointly brought all natural objects and processes as subjects of scientific investigation. There is no more reason to consider that Darwin's theory of evolution and explanation of design is anti-Christian than there is any reason to consider anti-Christian Newton's laws of motion. Divine action in the universe must be sought in ways other than those that postulate it as the means to account for gaps in the scientific account of the workings of the universe.

INTELLIGENT DESIGN'S VACUITY

The argument-from-design advanced by William Paley is a two-tined argument. The first prong asserts that humans, as well as all sorts of organisms in their wholes, in their parts, and in their relations to one another and to their environment, appear to have been designed for serving certain functions and for certain ways of life. The second prong of the argument is that only an omnipotent Creator could account for the perfection and functional design of living organisms. In the 1990s, several authors, notably biochemist Michael Behe (*Darwin's Black Box,* 1996), theorist William Dembski (*The Design Inference,* 1995), and law professor Phillip Johnson (*The Wedge of Truth,* 2000), among others, revived the argument-from-design. However, these authors sought to hide their real agendas, and thus, typically avoided explicit reference to God, so that the "theory" of intelligent design (ID) could be taught in the public schools, as an alternative to the theory of evolution, without incurring conflict with the U.S. Constitution, which forbids the endorsement of any religious beliefs in public institutions. The folly of the pretense that the ID argument is scientific rather than religious is apparent to anyone who takes the time to consider the issue seriously. I will not pursue this issue here because it has been repeatedly exposed. (See, for example, Chapter 8 of my *Darwin's Gift to Science and Religion,* 2007.)

Proponents of ID call for an Intelligent Designer to explain the supposed irreducible complexity in organisms. Behe defines an irreducibly complex system as an entity "composed of several well-matched, interacting parts that contribute to the basic function, wherein the removal of any one of the parts causes the system to effectively cease functioning" (1996, 39). They argue that irreducibly complex systems cannot be the outcome of evolution. According to Behe:

> An irreducibly complex system cannot be produced directly . . . by slight, successive modifications of a precursor system, because any precursor to an irreducible complex system that is missing a part is by definition nonfunctional. . . . Since natural selection can only choose systems that are already working, then if a biological system cannot be produced gradually it would have to arise as an integrated unit, in one fell swoop, for natural selection to have anything to act on. (1996, 72)

In other words, unless all parts of the eye come simultaneously into existence, the eye cannot function; it does not benefit a precursor organism to have just a retina or a lens if the other parts are lacking. The human eye, according to this argument, could not have evolved one small step at a time, in the piecemeal manner by which natural selection works. But evolutionists have pointed out again and again with supporting evidence that organs and other components of living beings are not irreducibly complex—they do not come about suddenly or in one fell swoop. Evolutionists have shown that organs such as the human eye are not irreducible at all; rather, less-complex versions of the same systems have existed in the past and can be found in today's organisms.

Eyes evolved gradually and achieved very different configurations in different organisms, all serving the function of vision. The simplest "organ" of vision occurs in some single-celled organisms that have enzymes or spots sensitive to light, which help them move toward the surface of their pond where they feed on the algae growing there. Some multicellular animals exhibit light-sensitive spots on their epidermis. Further steps—deposition of pigment around the spot, configuration of cells into a cuplike shape, thickening of the epidermis leading to the development of a cornea, development of muscles to move the eyes and nerves to transmit optical signals to the brain—gradually led to the highly developed eyes of vertebrates and cephalopods (octopuses and squids) and to the compound eyes of insects.

A record of the major stages in the evolution of a complex eye has survived in living mollusks (clams, snails, and squids). The eye of the octopus and squid is as complex as the human eye, with cornea, iris, refractive lens, retina, vitreous internal substance, optic nerve, and muscle. Limpets (*Patella*) have just about the simplest imaginable eye: just an eye spot consisting of a few pigmented cells with nerve fibers attached to them. Several intermediate stages are found in other living mollusks. One step in complexity above the limpet eye is found in slit-shell mollusks (*Pleurotomaria*), which have a cup-eye, one layer of pigmented cells curved like a cup with a wide opening through which light enters, with each pigmented cell in the back of the cup attached to a nerve fiber. More complex is the pinhole-lens eye found in Nautilus, a marine snail. The layer of pigmented cells is considerably more extensive than in slit-shell mollusks; the pigmented cells are covered toward the front with epithelium (skin) cells that are nearly closed except for a small opening ("pinhole") for passage of light, creating a cavity filled with water. Murex, another marine snail, has an eye with a primitive, refractive lens covered with epithelium cells (serving as a primitive cornea); the pigmented cells extend through the back of the eye cavity (thus serving as a retina) and the nerve fibers are collected into an optic nerve that goes to the brain. The most advanced mollusk eye is found in the octopus and the squid, which is just as complex and effective as the human eye and lacks the human blind spot, an imperfection due to the fact that the nerve fibers of the human eye are collected inside the eye cavity so that the optic nerve has to cross the retina on its way to the brain; the nerve fibers and the optic nerve of the octopus eye are outside the eye cavity and travel to the brain without crossing the retina.

The gradual process of natural selection adapting organs to functions occurs in a variety of ways, reflecting the haphazard characteristics of the evolutionary process, which are due to mutation, past history, and the vagaries of environments. In some cases the changes of an organ amount to a shift of function, as in the evolution of the forelimbs of vertebrates, which first evolved in reptiles originally adapted for walking, but which are now used in birds for flying, in whales for swimming, and in humans for handling objects. Other cases, as the evolution of eyes, exemplify gradual advancement of the same function—seeing. In all cases, however, the process is impelled by natural selection's favoring through time of individuals exhibiting functional advantages over others of the same species.

Examples of functional shifts are many and diverse. Some transitions at first may seem unlikely because of the difficulty in identifying which possible functions may have been served during the intermediate stages. These cases are eventually resolved with further research and often by the discovery of intermediate fossil forms or living organisms with intermediate stages of development, as in the case of mollusks' eyes. A favorite ID example of alleged irreducible complexity is the bacterial flagellum. The bacterial flagellum is, according to Behe, irreducibly complex because it consists of several parts so that, if any part is missing, the flagellum will not function. It could not, therefore, says Behe, have evolved gradually, one part at a time, because the function belongs to the whole, the separate parts cannot function by themselves. But this is not so. In different species of bacteria, there are different kinds of flagella, some simpler than the one described by Behe, others just different, even very different, as in the archaea, a very large bacteria-like group of organisms. Moreover, motility in many bacteria is accomplished without flagella at all. Still more, biochemists have shown that some flagellum components may have evolved from secretory systems, which are very similar to the flagellum but lack some of the flagellum's components.

The bacterial flagellum is not irreducibly complex: a subset of the flagellum's complement of proteins evolved as a mechanism for bacteria to inject proteins across a cell's membrane. The argument for the irreducible complexity of the flagellum is formulated, like other ID arguments, as an "argument-from-ignorance." Because one author does not know how a complex organ may have come about, it must be the case that it is irreducibly complex. This argument-from-ignorance dissolves as scientific knowledge advances or when pre-existing scientific knowledge is taken into account. This has, indeed, come about in the case of the bacterial flagellum. Two recent papers by M. J. Pallen and N. J. Matzke (2006) and by R. Liu and H. Ochman (2007) have delivered the final coup de grâce. The components of the flagellum are encoded by gene clusters that may include, in some species, upwards of fifty genes. Liu and Ochman have identified all the flagellar proteins in forty-one species from eleven quite diverse groups of bacteria. Twenty-four of the genes encoding the flagellar proteins were already present in the remote common ancestor of all the bacterial species studied. The other genes have come about by duplication and evolution of pre-existing genes. Moreover, many of the core twenty-four ancestral genes are also derived from a few pre-existing ones by successive gene duplications that gradually increased their number.

Other extravagant claims of Behe refer to the blood-clotting mechanism and to the immune system in mammals and other vertebrates. Astonishingly, Behe has written that "There is no publication in the scientific literature—in prestigious journals, specialty journals, or books—that describes how molecular evolution of any real, complex, biochemical system either did occur or even might have occurred," and in particular, "the scientific literature has no answers to the origin of the immune system" (1996, 185; 138). The vacuity of this claim was clearly stated by Federal Judge John E. Jones III in a judicial decision in December 2005, *Kitzmiller v. Dover Area School District,* which culminated in a several-weeks-long trial in Pennsylvania concerning the teaching of ID in the public schools. Judge Jones points out with understated disbelief that

> Professor Behe was questioned concerning his 1996 claim that science would never find an evolutionary explanation for the immune system. He was presented with fifty-eight peer-reviewed publications, nine books, and several immunology textbook chapters about the evolution of the immune system; however, he simply insisted that this was still not sufficient evidence of evolution, and that it was not "good enough." (78)

Judge Jones concludes: "We therefore find that Professor Behe's claim for irreducible complexity has been refuted in peer-reviewed research papers and has been rejected by the scientific community at large." The arguments for the irreducible complexity of the blood-clotting cascade and of the immune system are formulated, like the ID argument for the flagellum, as arguments from ignorance. As I have pointed out, the argument-from-ignorance dissolves as scientific knowledge advances, or when pre-existing scientific knowledge is taken into account.

INTELLIGENT DESIGN'S BLASPHEMY

Intelligent Design is bad science or not science at all. It is not supported by experiments, observations, or results published in peer-reviewed scientific journals. I further argue that ID is bad religion and bad theology because it implies that the designer has undesirable attributes that we don't want to predicate of God. It is not only that organisms and their parts are less than perfect, but also that deficiencies and dysfunctions are pervasive, evidencing "incompetent" rather than "intelligent" design. Consider the human jaw. We have too many teeth for the jaw's size so that wisdom teeth need to be removed and orthodontists make a decent living straightening the others. Do the IDers want to blame God for this blunder? A human engineer would have done better.

Evolution gives a good account of this imperfection. Brain size increased over time in our ancestors; the remodeling of the skull to fit the larger brain entailed a reduction of the jaw so that the head of the newborn would not be too large to pass through the mother's birth canal. Evolution responds to organisms' needs through natural selection, not by optimal design but by "tinkering," by slowly modifying existing structures. Evolution achieves "design" as a consequence of natural selection while promoting adaptation. Evolution's ID is *Imperfect* Design, not Intelligent Design.

Consider also the birth canal of women, much too narrow for easy passage of the infant's head, so many mothers and babies, thousands upon thousands, die during delivery, about 350,000 babies per year according to current estimates. Surely we don't want to blame God for this dysfunctional design or for the children's deaths. Science makes it understandable, a consequence of the evolutionary enlargement of our brain. Females of other primates do not experience this difficulty. Theologians in the past struggled with the issue of dysfunction because they thought it had to be attributed to God's design. Science, much to the relief of theologians, provides an explanation that convincingly attributes defects, deformities, and dysfunctions to natural causes.

One more human example: why are our arms and our legs, which are used for such different functions, made of the same materials, the same bones, muscles, and nerves, all arranged in the same overall pattern? Evolution makes sense of the anomaly. Our remote ancestors' forelimbs were legs. After our hominid ancestors became bipedal and started using their forelimbs for functions other than walking, the forelimbs became gradually modified, but retained their original composition and arrangement. Engineers start with raw materials and a design suited for a particular purpose; evolution can only modify what is already there. An engineer who would design cars and airplanes or wings and wheels, using the same materials arranged in a similar pattern, would surely be fired.

Examples of deficiencies and dysfunctions in all sorts of organisms can be endlessly multiplied, reflecting the opportunistic, tinkerer-like character of natural selection, which achieves imperfect design rather than intelligent design. The world of organisms also abounds in characteristics that might be called "oddities," as well as those that have been characterized as "cruelties," an apposite qualifier if the cruel behaviors were designed outcomes of a being holding on to human or higher standards of morality. But the cruelties of biological nature are only metaphoric cruelties when applied to the outcomes of natural selection.

Examples of "cruelty" involve not only the familiar predators (say, a chimpanzee) tearing apart their prey (say, a small monkey held alive by a chimpanzee biting large flesh morsels from the screaming monkey) or parasites destroying the functional organs of their hosts, but also, and very abundantly, between organisms of the same species, including between mates. A well-known example is the female praying mantis that devours the male after coitus is completed. In some midges (tiny flies), the female captures the male as if he were any other prey, and with the tip of her proboscis, she injects into him her spittle, which starts digesting the male's innards that are then sucked by the female; partly protected from digestion are the relatively intact male organs that break off inside the female and fertilize her. Male cannibalism by their female mates is known in dozens of species, particularly spiders and scorpions. The world of life abounds in "cruel" behaviors. Numerous predators eat their prey alive; parasites destroy their living hosts from within; in many species of spiders and insects, as noted, the females devour their mates.

Religious scholars in the past had struggled with imperfection, dysfunction, and cruelty in the living world, which are difficult to explain if they are the outcomes of God's design. The philosopher David Hume set the problem succinctly with brutal directedness: "Is he [God] willing to prevent evil, but not able? Then he is impotent. Is he able, but not willing? Then he is malevolent. Is he both able and willing? Whence then evil?" (1907, 134). Evolution came to the rescue. John Haught, a contemporary Roman Catholic theologian, has written of "Darwin's gift to theology" (1998). The Protestant theologian Arthur Peacocke (2001) has referred to Darwin as the "disguised friend," by quoting the earlier theologian Aubrey Moore, who in 1891 wrote that "Darwinism appeared, and, under the guise of a foe, did the work of a friend." Haught and Peacocke are acknowledging the irony that the theory of evolution, which at first had seemed to remove the need for God in the world, now has convincingly removed the need to explain the world's imperfections as outcomes of God's design.

Indeed, a major burden was removed from the shoulders of believers when convincing evidence was advanced that the design of organisms need not be attributed to the immediate agency of the Creator, but is rather an outcome of natural processes. If we claim that organisms and their parts have been specifically designed by God, we have to account for the incompetent design of the human jaw, the narrowness of the birth canal, and our poorly designed backbone, which is less than fittingly suited for walking upright. Proponents of intelligent design would do well to acknowledge Darwin's Revolution and accept natural selection as the process that accounts for the design of organisms, as well as for the dysfunctions, oddities, cruelties, and sadism that pervade the world of life. Attributing these to specific agency by the Creator amounts to blasphemy, an implication of their ideas when one takes into account biological knowledge.

Most disturbing is the following consideration, which amounts to a monumental insult to the Creator by people of faith who believe that the biology of human beings was specifically designed by God. About 20 percent of all human pregnancies end in spontaneous abortion during the first two months, owing to the deficiencies of the human reproductive system. ID implicitly attributes this calamity to the Creator's faulty design, rather than to the clumsy ways of the evolutionary process. There can be no escape from the conclusion that Intelligent Design is blasphemy.

THE POWER AND LIMITS OF SCIENCE

Science is a wondrously successful way of knowing. Science seeks explanations of the natural world by formulating explanations based on observation and experimentation that are subject to the possibility of rejection or corroboration by cycles upon cycles of additional observations and experimentation. A scientific explanation is tested by ascertaining whether or not predictions about the world of experience derived from the explanation agree with what is later observed.

Science as a mode of inquiry into the nature of the universe has been successful and of great consequence. Witness the proliferation of science academic departments in universities and other research institutions, the enormous budgets that the body politic and the private sector willingly commit to scientific research, and its economic impact. The Office of Management and Budget (OMB) of the U.S. government has estimated that 50 percent of all economic growth in the United States since the Second World War can be directly attributed to scientific knowledge and technical advances. The technology derived from scientific knowledge pervades our lives: the high-rise buildings of our cities, thruways and long span-bridges, rockets that bring men to the moon, telephones that provide instant communication across continents, computers that perform complex calculations in millionths of a second, vaccines and drugs that keep bacterial parasites at bay, and gene therapies that replace DNA in defective cells. All these remarkable achievements bear witness to the validity of the scientific knowledge from which they originated.

Scientific knowledge is also remarkable in the way it emerges by way of consensus and agreement among scientists, and in the way new knowledge builds upon past accomplishments rather than starting anew with each generation or each new practitioner. Surely scientists disagree with each other on many matters; but these are issues not yet settled, and the points of disagreement generally do not bring into question previous knowledge. Modern scientists do not challenge that atoms exist, that there is a universe with a myriad of stars, that heredity is encased in DNA, or that organisms have evolved from remote ancestors much different from them.

Science is a way of knowing, but it is not the only way. Knowledge also derives from other sources, such as common sense, artistic and religious experience, and philosophical reflection. The validity of the knowledge acquired by non-scientific modes of inquiry can be simply established by pointing out that science (in the modern sense of the word) dawned in the sixteenth century, but mankind had for centuries built cities and roads, brought forth political institutions and sophisticated codes of law, advanced profound philosophies and value systems, and created magnificent plastic art, as well as music and literature. We thus learn about ourselves and about the world in which we live, and we also benefit from products of this non-scientific knowledge. We learn about the human predicament reading Shakespeare's *King Lear,* viewing Rembrandt's self portraits, and listening to Tchaikovsky's *Symphonie Pathétique* or Elton John's "Candle in the Wind." The crops we harvest and the animals we husband emerged millennia before science's dawn, from practices set down by farmers in the Middle East, Andean sierras, and Mayan plateaus.

It is not my intention to belabor the extraordinary fruits of non-scientific modes of inquiry. But I have set forth the view that nothing in the world of nature escapes the scientific mode of knowledge and that we owe this universality to Darwin's Revolution. Here I wish simply to state something that is obvious but becomes at times clouded by thoughtlessness or hubris. Successful as it is, and universally encompassing as its subject is, a scientific view of the world is hopelessly incomplete. There are matters of value, meaning, and purpose that are outside science's scope. Even when we have a satisfying scientific understanding of a natural object or process, we are still missing matters that may well be thought by many to be of equal or greater import. Scientific knowledge may enrich aesthetic and moral perceptions and illuminate the significance of life and the world, but these are matters outside science's realm.

CODA

On April 28, 1937, early in the Spanish Civil War, Nazi airplanes under Franco's command bombed the small Basque town of Guernica, the spiritual home of the Basques, killing 1,654 of its 7,000 inhabitants, the first time that a civilian population had been determinedly destroyed from the air. The Spanish painter Pablo Picasso had recently been commissioned by the Spanish Republican Government to paint a large composition for the Spanish pavilion at the Paris World Exhibition of 1937. In a frenzy of manic energy, the enraged Picasso sketched in two days and fully outlined in ten more days his famous *Guernica,* an immense painting of 25 feet, 8 inches by 11 feet, 6 inches.

Suppose that I now would describe the images represented in the painting, their size and position, as well as the pigments used and the quality of the canvas. This description would be of interest, but it would hardly be satisfying if I had completely omitted aesthetic appreciation and considerations of meaning, the dramatic message of man's inhumanity to man conveyed by the outstretched figure of the mother pulling her killed baby, the bellowing faces, the wounded horse, or the satanic image of the bull. The point is that the physical description of the painting does not tell us anything (by itself it cannot tell us anything) about the aesthetic value or historical significance of *Guernica,* nor, on the other hand, do aesthetic or intended meaning determine the physical features of the painting.

Let *Guernica* be a metaphor of the final point I wish to make. Scientific knowledge, like the description of the size, materials, and geometry of *Guernica,* is satisfying and useful. But once science has had its say, there remains much about reality that is of interest—questions of value, meaning, and purpose that are forever beyond science's scope.

NOTE ON THE COMPATIBILITY OF A CREATED WORLD WITH THE PRESENCE OF EVIL

An argument could be advanced that the process of evolution by natural selection does not discharge God's responsibility for the dysfunctions, cruelties, and sadism of the living world because for people of faith, God is the creator of the universe and thus would be accountable for its consequences, direct or indirect, immediate or mediated. God, the argument would say, could have created a world where such things as cruelty, parasitism, and human miscarriages would not occur.

One possible answer is to claim that God's deeds are inscrutable and humans are not entitled to seek understanding of God's purposes, much less to bring His actions into account. This answer may seem to many unsatisfactory because it simply evades the question instead of answering it.

Other answers are, however, possible. One explanation that may be acceptable to some religious believers, but perhaps not all, would go along the following lines of reasoning. Consider first human beings, who perpetrate all sorts of misdeeds and sins, even perjury, adultery, and murder. People of faith believe that each human being is a creation of God, but this does not entail that God is responsible for human crimes and misdemeanors. Sin is a consequence of free will; the flip side of sin is virtue. Christian theologians have expounded that if humans are to enter into a genuinely personal relationship with their maker, they must first experience some degree of freedom. The eternal reward of heaven calls for a virtuous life. The critics might say that this account does not excuse God because God could have created "humans" (whatever they may have been called and been like) without free will. But one could reasonably argue that "humans" without free will would be a very different kind of creature, much less interesting and creative. Robots are not a good replacement for humans; robots do not perform virtuous deeds.

Before modern physical science came about, God (in some religious views) caused rain, drought, volcanic eruptions, and so on to reward or punish people. This view entails that God would have caused the tsunami that killed 200,000 Indonesians a few years ago. That would seem incompatible with a benevolent God. However, we now know that tsunamis and other "natural" catastrophes come about by natural processes. Natural processes don't entail moral values. Some critics might say "that does not excuse God, because God created the world as it is. God could have created a different world, without catastrophes." Yes, according to some belief systems, God could have created a different world. But that would not be a creative universe where galaxies form, where stars and planetary systems come about, and where continents drift and cause earthquakes. And the world that we have is creative and more exciting than a static world. This argument will not convince all, but is a valid argument for some as an account of physical evil.

Turn now to badly designed human jaws, parasites designed to kill millions of children, and a poorly designed human reproductive system. If these dreadful happenings are direct designs by God, God would seem responsible for the consequences. If engineers design cars that explode after 100 miles of driving, they are accountable. But if the dreadful happenings come about by natural processes (evolution), there are no moral implications because natural processes don't entail moral values. But some would say: the world was created by God, so God is ultimately responsible; God could have created a world without parasites or dysfunctionalities. Yes, others would answer, but a world of life with evolution is much more exciting; it is a creative world where new species arise, complex ecosystems come about, and humans have evolved.

The account that I have sketched will not satisfy some people of faith and many unbelievers will surely find it less than cogent—a deus ex machina. But I am suggesting that it may provide the beginning of an explanation for multiple others.

NOTE ON REFERENCES

The literature on the perceived conflict between evolution and intelligent design is very extensive. A few helpful books are the following

Ayala, Francisco J. 2006. *Darwin and Intelligent Design.* Minneapolis, MN: Fortress Press.

Dembski, William A., and Michael Ruse, eds. 2004. *Debating Design: From Darwin to DNA.* Cambridge, UK: Cambridge University Press.

Forrest, Barbara, and Paul R. Gross. 2004. *Creationism's Trojan Horse. The Wedge of Intelligent Design.* New York: Oxford University Press.

Kitcher, Philip. 2007. *Living with Darwin: Evolution, Design and the Future of Faith.* Oxford, UK: Oxford University Press.

Matsumura, Molleen, ed. 1995. *Voices for Evolution.* Berkeley, CA: National Center for Science Education.

Miller, Kenneth R. 1999. *Finding Darwin's God.* New York: Harper Collins.

Pennock, Robert T. 2002. *Tower of Babel: The Evidence against the New Creationism.* Cambridge, MA: MIT Press.

Petto, Andrew J., and Laurie R. Godfrey, eds. 2007. *Scientists Confront Intelligent Design and Creationism.* New York: W. W. Norton.

Ruse, Michael. 2005. *The Evolution-Creation Struggle.* Cambridge, MA: Harvard University Press.

Russell, Robert J., William R. Stoeger, and Francisco J. Ayala, eds. 1998. *Evolutionary and Molecular Biology: Scientific Perspectives on Divine Action.* Vatican City State and Berkeley, CA: Vatican Observatory and the Center for Theology and the Natural Sciences.

Scott, Eugenie C. 2004. *Evolution vs. Creationism: An Introduction.* Westport, CT: Greenwood Press.

Young, Matt, and Taner Edis, eds. 2004. *Why Intelligent Design Fails.* New Brunswick, NJ: Rutgers University Press.

See also: The extensive judicial decision of December 20, 2005 by Judge John E. Jones III is an articulate critique of the intelligent design movement and its pretense of being science-based rather than religious: *Kitzmiller v. Dover Area School District,* 400 F. Supp. 2d 707 (M.D. Pa. 2005), available at http://www.pamd.uscourts.gov/kitzmiller/kitzmiller_342.pdf.

REFERENCES

Ayala, Francisco J. 2007. *Darwin's Gift to Science and Religion.* Washington, DC: Joseph Henry Press.

Behe, Michael. 1996. *Darwin's Black Box: The Biochemical Challenge to Evolution.* New York: The Free Press.

Bell, Charles. 1833. *The Hand, its Mechanisms and Vital Endowments, as Evincing Design.* Philadelphia, PA: Carey, Lea & Blanchard.

Cannon, Walter F. 1961. "The Impact of Uniformitarianism: Two Letters from John Herschel to Charles Lyell, 1836–1837." *Proceedings of the American Philosophical Society* 105(2): 301–14.

Darwin, Charles. 1859. *On the Origin of Species by Means of Natural Selection.* London: John Murray.

———. 1958. *The Autobiography of Charles Darwin: 1809–1882.* Edited by Nora Barlow. London: Collins.

———. 1967. *The Origin of Species.* Facsimile of the first edition. New York: Atheneum.

———. 1991. "Letter to Charles Lyell, 18 June 1858." In *The Correspondence of Charles Darwin, Volume 7,* edited by Frederick Burkhardt and Sydney Smith, 107–8. Cambridge, UK: Cambridge University Press.

Dembski, William. 1995. *The Design Inference: Eliminating Chance through Small Probabilities.* Cambridge, UK: Cambridge University Press.

Haught, John F. 1998. "Darwin's Gift to Theology." In *Evolutionary and Molecular Biology: Scientific Perspectives on Divine Action,* edited by Robert J. Russell, William R. Stoeger, S. J. Francisco, and J. Ayala, 393-418. Vatican City State and Berkeley, CA: Vatican Observatory Publications and Center for Theology and the Natural Sciences.

Hoving, Thomas. 1993. *Making the Mummies Dance: Inside the Metropolitan Museum of Art.* New York: Simon & Schuster.

Hume, David. 1907. *Dialogues Concerning Natural Religion.* Originally published 1779. London: William Blackwood and Sons.

John Paul II. 1996. Message to Pontifical Academy of Sciences. Translated by *L'Osservatore Romano,* "Weekly Edition in English," 30 October.

Johnson, Phillip. 2000. *The Wedge of Truth.* Downers Grove, IL: InterVarsity Press.

Kitzmiller v. Dover Area School District. 400 F. Supp. 2d 707 (M.D. Pa. 2005), available at http://www.pamd.uscourts.gov/kitzmiller/kitzmiller_342.pdf.

Liu, Renyi, and Howard Ochman. 2007. "Stepwise Formation of the Bacterial Flagellar System." *Proceedings of the National Academy of Sciences USA* 104: 7116–21.

Paley, William. 1802. *Natural Theology.* London: Wilks and Taylor.

Pallen, Mark J., and Nicholas J. Matzke. 2006. "From the *Origin of Species* to the Origin of Bacterial Flagella." *Nature Reviews Microbiology* 4: 784–90.

Peacocke, Arthur. 2001. "Welcoming the 'Disguised Friend'—Darwinism and Divinity." In *Intelligent Design Creationism and Its Critics: Philosophical, Theological, and Scientific Perspectives,* edited by Robert T. Pennock, 471–86. Cambridge, MA: MIT.

Strong, Augustus H. 1886. *Systematic Theology.* Rochester, NY: E. R. Andrews.

Chapter 2

Creating New Paradigms in Political Science

The Case of Biology and Politics

Rose McDermott and Kristen Renwick Monroe

Conversation

Kristen Monroe (KM): Let me begin by asking how you got interested in the biological approach to politics, in terms of your own career.

Rose McDermott (RM): I think I came to it accidentally, as an extension of the original work that I had done in political psychology. In many ways, my impetus for moving in this direction came from Steve Rosen, who obtained funding from Andy Marshall at the Office of Net Assessment within the Department of Defense. Steve Rosen is a professor of political science at Harvard in the government department. He started out as an India specialist. He became very interested in the brain. He had an ongoing relationship with Andy Marshall, who is the head of the Office of Net Assessment in the Department of Defense. Steve Rosen set up a lunch when I was at Harvard on a fellowship from 2001 to 2002. I was in the Kennedy School Women and Public Policy Program Fellowship.

I had lunch with Andy Marshall and Steve Rosen, and I told them about a series of experiments that I had conducted at Cornell looking at gender differences in aggression. I explained how the original experiment was really nothing to do with gender—it was really about how you make judgments about uncertainty. I had a war game scenario about how people would make judgments about uncertainty. One of my Research Assistants came in one day very early on in the experiment and said, "You know the girls are just really different than the boys." I thought, "Hmm. Okay, do you think we can go back and figure out which ones were the girls and which ones were the boys?" She said, "Oh yeah, you can tell by the handwriting." So they went back and figured out past subjects and thereafter coded whether they were women or men. But the first ten subjects I ran, I didn't even code gender. I was *so* not thinking in that direction. But the only robust results that came out of that study were gender. I didn't get anything on uncertainty.

I was explaining this to Andy Marshall and, as I remember the discussion, he said, "We have a real problem in the military because we can't get women to share food with men in training. We send out a group of men and they all get along and share food and it's wonderful. They come back and they're all bonded. But you send out groups of women with men, and the women won't share food. The men and women do not bond well under these conditions." This was a real problem for morale and unit cohesion. Then Andy asked me, "Can you help me figure out what's really going on here?" He very generously offered three-year funding from 2001 to 2004. This helped fund, in part, the research for the book I did on presidential decision making in foreign policy[1] and a series of experiments on gender differences in aggression.

In the course of doing that study, I became extremely interested in learning about hormones and their impact on decision making and judgment and so on. These experiments resulted in several publications, one of which was on the effect of positive illusions and overconfidence in conflict, published in the *Proceedings of the Royal Society of London*.[2] That was my first publication in a top-five hard-science journal. The impact of those journals is really different. The piece on testosterone came out in *The Annals* in November 2007.[3] There's a piece on communications effects that came out in the *Journal of Peace and Conflict* last summer.[4] And there was a piece on finger length. Emily Barrett, from the biological anthropology department at Harvard, did the testosterone essays and she had a friend, Matthew McIntyre, who was very interested in finger length ratio to conflict and that piece came out in *Personality and Individual Differences*.[5] The person who ended up doing the statistics was Dominic Johnson at Edinborough. Dominic was very interested in genetic influences because he had a five-year fellowship from this organization in Switzerland—Branco-Weiss. Independently, both he and I had been thinking about the possible influence of MAO (monoamine oxidase) on aggression. In addition, there was an Italian geneticist—Giovanni Frazzetto—who was part of the group, and so the three of us got together and did this MAO study, which was published in the *Proceedings of the National Academy of Sciences*.[6] We are still looking at some additional variables we collected on public opinion measures and confidence.

KM: So, in your case, the development of your interest in biological approaches to the study of politics was chance?

RM: Yes, absolutely. Accidental.

KM: But is it fair to say you also saw a possibility, and you were willing to consider something outside the mainstream?

RM: Yes. I think that's because I've never been in the mainstream. It was easier, because I wasn't going to have to pay a professional cost in the sense of giving up being front and center. I think a lot of it also had to do with ongoing relationships I had with people who had other, but related, interests and were in positions to provide something.

KM: But also, there's a kind of receptivity to innovative and different work that's there in your own mind. Why is it that you think that some people have it and others don't?

RM: That's a tough one. It's really an interesting question. In my case, some part of it really was training. When I was an undergrad, I was trained in both the political science and the psychology departments. Now, some of that was accident, too. One quarter, I was looking for a class and the time of Amos Tversky's judgment and decision-making class fit my schedule. I walked in the room. I still remember it like it was yesterday. He stood up in front of the room, five feet, ninety pounds, and you knew that he had killed people with his bare hands. He was just a force of nature. He stood up there and said, "There's one existential problem. No one else can take a shower for you. Everything else is about understanding how that relates to how you think about things." I thought, "Wow!" I was young enough to think that was deep! I was just captivated. It was the most intellectually enthralling experience I'd ever had in my life. That was my sophomore year of college, but I knew that's what I was going to write my dissertation on, and that indeed is what I wrote my dissertation on. So some of it really was about training and a lot of it is about mentorship, people who encourage you, who tell you that this is a good idea and that you can do it and that they'll help you do it.

KM: You're into what I would call the biological approach to politics. I wonder how you would define that field in general. Then maybe we could talk a little bit about the history of that field in political sciences because—and please correct me if I'm wrong—it seems to me that there was a movement before, and the biological movement failed. Now, I'm seeing a resurgence coming through the kind of neuroscience work that's being done.

RM: Okay. But first I'd like to just go back to one big issue that I ended that question on, which is your issue about ego-strength. I think to me a lot of it's really about appreciation. A lot of us are in academics because we want other people to appreciate our work and appreciate what we do. And yet, what everybody does is just expect everybody else to appreciate them without expecting to appreciate anybody, and so to me it's a lot about reciprocity in that regard. One of the most powerful things that someone can do is, in a meaningful way, to show that you appreciate the effort that somebody else has put in. A lot of people in political science—and this is less true in psychology—consider it a weakness to acknowledge that somebody else has done good work, because somehow, it makes your work less valuable or something.

KM: They treat academia as a zero-sum game.

RM: Yes. I think that's wrong. What that does is prevent the kind of social relationships that can build into interdisciplinary work.

Now, about biology and politics: I guess I wouldn't necessarily define it or think about it as biology and politics. I really think of it more as a kind of combination of cognitive neurosciences, behavior genetics and politics. The reason I do that is because I think that the old biology and politics movement is exactly as you describe. There were two problems with that work. One was, for whatever set of reasons—and I don't think it was their fault—but the movement got associated with racism. It got impaled on this notion of Social Darwinism. A lot of the work had that element, although certainly not all of it, but the association was there in the minds of many observers, whether justified or not. I think that association is what caused the old biology and politics movement to die. People didn't want to be associated with work that looked racist and sexist, and had these kinds of hierarchies that were politically incorrect and inappropriate and morally wrong. That response against that kind of work was correct. That kind of work should die because of that.

The other problem was that what a lot of that work actually did was issue a call to do the work without actually doing the work itself. So it's as if they said, "We should do biology and politics." But nobody actually sat down and drilled down the genetic models and took the hormonal assays and actually did that science. Instead, they just read the work of others in the hard sciences and speculated on what that meant for politics. And a lot of this application was, and continues in some ways to be, based on a misunderstanding of the basic science. No real geneticist, for example, would ever talk about something as simplistic as "the gene for" any particular outcome outside an excruciatingly few dominant genetic diseases like Huntington's. You need people who don't just say, "Other people should really do this, this is really important." You have to get down there in the trenches and do the work yourself.

KM: Okay, though is it also perhaps a question of intellectual propinquity? Political scientists are over here and biology is over there.

RM: That's what I mean by alliances, right? If I have a social network with a psychologist and that psychologist has a social network with a geneticist, I'm two steps removed. I don't have to understand the genetics; I just have to trust my psychological ally. The psychological ally doesn't need to understand the politics. Neither does the geneticist. You just have to make sure you trust the credibility of your allies. I think that is what's different this time—and I would date that change from, let's call it 1990 for a lack of a better date—there's a real emerging consensus that's beginning to happen, outside of political science, in a lot of other sciences. Biology, cognitive neuroscience, psychology, ecology, and a number of these other areas, like behavior genetics, all are coming together around two things. One is these phenomenal, just catastrophically fast, findings resulting from new technologies like MRIs that really change our ability to think about things. It's not just MRIs, it's also the human genome project and being able to do hormonal assays at an affordable rate on two hundred people and being able to combine those things. So it's very fast scientific technological developments in more than one field. Those things would've stayed separate if it weren't for the coming together of a kind of consensus around a particular theoretical model, which is evolutionary design and development. So to the extent that all these groups come together, and think about adaptation, think about functionality, think about by-products in the same way, it allows for different kinds of conversations across these fields.

It's like a common language. Before, you had a language called biology and you had a language called genetics and you had a language called psychology and you had translators. But now we have Esperanto. We have one language. It's not the case anymore that geneticists say to psychologists, "You don't understand what I'm doing," because they have a joint model. They have a joint argot. It's like David Brooks, from the *New York Times* saying, "Evolution is the meta-narrative of our time." I think that's how he means it. It is not that nobody believes in God or that everyone believes that evolution is why everything is the way it is. What I mean is that the conversation by which people understand their process across disciplines is converging around evolutionary models. Evolution represents a very hard-working theory; it can generate a lot of robust and powerful explanatory hypotheses that can then be tested in a variety of ways using diverse tools. How do we think about how people came to have genetic structure? How do we think about how people came to have this kind of brain architecture? What purpose might it serve? What function does it accomplish? How do we come to think about the way that people have these kinds of hormonal systems? How do these things interact? Because they have the same model, they can have the same conversation.

KM: Earlier we talked about the "zero-sum" game. What I'm hearing you say—correct me if I'm wrong—is that the people who do interdisciplinary work are thinking more in terms of change and collaboration in a non-"zero-sum" game.

RM: Absolutely.

KM: And the people who are still thinking in terms of a zero-sum game, people who say, "Well, I don't want to share my work with you, I don't want to put my graduate research assistants' names on my article"—that kind of mentality limits people who want to do interdisciplinary work and in some ways limits people who want to do creative scientific work.

RM: There's no question about it, and it holds scientific progress back. I think it is the primary and most fundamental unconscious and unexamined bias that keeps political science at an impasse. It shows up mostly in publications and promotion aspects of political science. If you're not the cowboy writing a single-author paper, you're not going to get tenure. I've seen important political scientists—we'll not mention names—tell people that that's what they advise graduate students. The problem is that that's not how real science gets done in other fields for two reasons. One is they like to work together. It's fun to collaborate; it's fine to have ten authors on a piece. But, second, a lot of this stuff is so technologically advanced that one person can't do all of it. So you need a geneticist, and a person who can do MRIs, and a statistician, and a theoretician, and you bring together your strengths, and then it really is the case that the sum is bigger than the collection of the parts.

KM: Of course, a lot of the hard scientists have competitive models still. They dump people from the lab if they don't go along with it. The organization of the collaboration can be a very hierarchical one.

RM: Absolutely. They very much have a kind of a structured mentorship. You have your advisors and you have your students. And your students take over from you, and there's this kind of way that it develops. But I mean, if you look at it that way—Duke just made a woman head of the medical school this week. It's the first top-ten medical school that's ever been headed by a woman. There's an acting head at the Harvard Medical School, but she's not a permanent one. It's 2007; I mean, if medicine can advance faster than political science, that's a sad statement.

KM: But now what you are talking about is collaborative work, not interdisciplinary work. Is interdisciplinary work most frequently done as a collaboration?

RM: I think so, because it's easier to have each person bring a strength from his or her field, and so it's much easier to accomplish good work interdisciplinarily if you have collaborators in those other disciplines. The problem I see is you don't get rewarded for that within your own discipline, right? So, I can have publications in hard science journals like *Proceedings of the Royal Society* and in *PNAS* (*Proceedings of the National Academy of Science*) and not get departmental credit even though the impact of these journals is much higher. Those are among the four or five most important scientific journals in the world, read by ten times the number of people that subscribe to the *American Political Science Review (APSR).* Yet by publishing there, I will not get as much promotion credit, I will not get as much pay credit, I will not get as much esteem by my colleagues as if I had an *APSR* article. But the difference is I can then walk into a psychology department or a biological anthropological department and they'll go, "Oh, you wrote that *PNAS* piece." So, it gives you credibility across disciplines in a different sort of way.

KM: So, we need to change the reward structure in political science?

RM: Absolutely!

KM: How do we do that?

RM: I'm very depressed about it. I've seen people sitting in conferences with twenty high-powered people saying, "I don't encourage my graduate students to do interdisciplinary work or collaborative work." I feel like there used to be such work, but there's less and less. I think part of it is that it gets harder and harder. I mean, you can see it within psychology itself. People used to do studies where they didn't use college students. But now it's just too hard to get enough work done because you have to do a certain number of studies in a certain number of years, and so it's more and more about less and less.

KM: And we're basing all these experiments on these college students. But think about it. We're basing all kinds of conclusions about politics on people for whom politics is much less important than sex.

RM: Right, right. But it's in the incentive structures of the institutions. You're supposed to do these big projects. But by that time you're old and tired, and it's very hard for someone at forty years old to say, "I'm going to take on a new thing. I'm going to learn how to do MRIs. I'm going to learn how to do hormonal analysis." The system rewards people with publications and grants, big grants. You don't get those by retooling and learning a new field.

KM: Do you think another problem in political science—for people doing biological work, anyway—is the fascination we have with behavioral data and quantitative work?

RM: Maybe. I don't do the quantitative, mainstream, large studies. If I sat around and did number crunching of the correlates of war, I'd be given an easier time.

KM: I wonder if that's related to the penchant for replicating the traditional paradigm, which works against creative work perhaps more than interdisciplinary work.

RM: Well, people just seek out other people who validate their beliefs. I think that is oftentimes the problem with established funding agencies like NSF. You know, they send these proposals to reviewers who are established people and who are very invested in their particular view of the world. It's common to get rejections from NSF that say some version of, "Oh they're going to do it anyways, so we don't need to fund it." Or "this is just a database and there's no theoretical contribution here." It's like there's a very particular kind of person they go to who is well established in the field, and the more well established they are, the more invested they're going to be in the existing wisdom.

KM: These people often edit the journals, too.

RM: It's interesting when people who get in charge of journals either haven't published a lot themselves or have a very committed ideological position and don't want to publish people who do anything different. I think it's what also makes interdisciplinary work hard. It's very hard to find journals that will accept more creative work, because if an editor wants to make a choice between a paper that's going to be the umpteenth piece on the democratic peace that validates all of his or her work, he or she is going to choose that over something that's going to threaten to undermine it. And this is not unique to political science, by any means. Most work in behavioral economics these days is simply variants on a theme of the ultimatum game and the dictator game and so on. As my friend Leda Cosmides says, you just have to keep in mind Max Planck's dictum that social science progresses one funeral at a time. You have to keep in mind that the people you have to convert are below you; they're not above you. The graduate students are the ones whose minds remain open, and also they are the ones you need to engage to change the nature of future research and ideas about how research should progress, both empirically and theoretically. Let them know that the world is broader than the six people that they've been told to read in graduate school.

KM: Again, this raises the question of how does one encourage good, new creative work—not necessarily interdisciplinary work, because theoretically you could get an entrenched interdisciplinary approach that had all the problems we're talking about. Indeed, the mathematical approach could be said to suffer from these problems.

RM: Yes. This is one of the things I talked to Leda about. Leda is kind of the mother of evolutionary psychology and really has as much grit and perseverance as anyone I've ever known. She's a genius. She said if you want to change a field, there are a few things you have to do. The first is you have to call it what it is. You've got to call it a revolution. Every time you talk to people, you've got to say, I'm at the forefront of the evolutionary psychology revolution and/or the cognitive neuroscience revolution, or the political ecology revolution. Call it what it is. Tell everyone you know. Try to make a self-conscious effort to change people's views.

She said the second thing you have to do is have a group. Some of them could be graduate students, some of them could be high-powered, but they join together over time and give one another ideas and support and collaboration. I consciously chose people outside of political science to join.

She said the third thing is you have to have a seminal text. It's critical. Yet it is increasingly difficult to get edited volumes done in political science. A lot of presses won't even look at them. They just turn them down. That is why I think it was a good idea for us to be able to co-edit the special mini-symposium in *Political Research Quarterly.*[7] It is the perfect thing because graduate students can cite it; it becomes a basic reference for people. And it gives permission to the next generation that this is legitimate work, that this is interesting work that high-powered people can do in a credible way, and that precipitates the snowball effect. It can grow.

So that's how I've been trying to advance. Leda said, you know, it's important to have a few high-visibility publications like the piece that I did on neuroscience in *Perspectives,* the piece that Hibbing and Alfred did on twin studies and the piece that they did on the origins of politics, and the piece that John Orbell did in *APSR.*[8] And then there are the great additional pieces that James Fowler has been doing with his graduate student Chris Dawes, and the super-impressive work that Pete Hatemi has been doing with Lindon Eaves.[9] There's now a small but increasing number of really good things by really good people in really good journals and that helps. If it's not a journal, it's a book—whatever it happens to be, but just to produce something that people can look at and say, "You can do this—great." Take these neuroeconomics people; they're in love because the MRI people have this great technology and no questions, and the economics people have the questions but no technology. But the problem is economics doesn't have the most interesting questions; the real interesting, broader questions are the political questions and the social questions. But political scientists tend to have even less access to advanced technology and less established collaborative relationships with people outside our own discipline. But we have great, important questions: How is it that we get along? Why don't we get along? How is it that we form hierarchies? How do we form friendships? Why do we have prevalent sex differences? All of these kinds of things. That's what needs to be investigated, and that's much more interesting and important, but nobody's doing it.

KM: Well, perhaps this interview will encourage scholars to work in such an area.

NOTES

1. McDermott 2007.
2. Johnson et al. 2006.
3. McDermott et al. 2007.
4. McDermott, Cowden, and Rosen 2008.
5. McIntyre et al. 2007.
6. McDermott et al. 2009.
7. McDermott 2009.
8. McDermott 2004; Alford, Funk, and Hibbing 2005; Alford and Hibbing 2004; Orbell et al. 2004.
9. See, e.g., Dawes and Fowler 2009; Eaves and Hatemi 2008.

REFERENCES

Alford, John R., Carolyn L. Funk, and John R. Hibbing. 2005. "Are Political Orientations Genetically Transmitted?" *American Political Science Review* 99(2): 153–67.

Alford, John R., and John R. Hibbing. 2004. "The Origin of Politics: An Evolutionary Theory of Political Behavior." *Perspectives on Politics* 2: 70723.

Dawes, Christopher T., and James H. Fowler. 2009. "Partisanship, Voting, and the Dopamine D2 Receptor Gene." *Journal of Politics* 71(3): 1157–71.

Eaves, Lindon J., and Peter K. Hatemi. 2008. "Transmission of Attitudes toward Abortion and Gay Rights: Effects of Genes, Social Learning and Mate Selection." *Behavior Genetics* 38(3): 247–56.

Johnson, Dominic D. P, Rose McDermott, Emily S. Barrett, Jonathan Cowden, Richard Wrangham, Matthew H. McIntyre, and Stephen P. Rosen. 2006. "Overconfidence in Wargames: Experimental Evidence on Expectations, Aggression, Gender and Testosterone." *Proceedings of the Royal Society of London B (Biological Science)* 273: 2513–20.

McDermott, Rose. 2004. "The Feeling of Rationality: The Meaning of Neuroscientific Advances for Political Science." *Perspectives on Politics* 2(4): 691–706.

———. 2007. *Presidential Illness, Leadership and Decision Making.* Cambridge, UK: Cambridge University Press.

———. 2009. "Mutual Interests: The Case for Increasing Dialogue between Political Science and Neuroscience." *Political Research Quarterly* 62(3): 571–83.

McDermott, Rose, Jonathan A. Cowden, and Stephen Rosen. 2008. "The Role of Hostile Communications in a Crisis Simulation Game." *Peace and Conflict* 14(2): 151–68.

McDermott, Rose, Dominic Johnson, Jonathan Cowden, and Stephen Rosen. 2007. "Testosterone and Aggression in a Simulated Crisis Game." *The ANNALS of the American Academy of Political and Social Science* 614: 15–33.

McDermott, Rose, Dustin Tingley, Jonathan Cowden, Giovanni Frazzetto, and Dominic D. P. Johnson. 2009. "Monoamine Oxidase A Gene (MAOA) Predicts Behavioral Aggression following Provocation." *Proceedings of the National Academy of Sciences* 106(7): 2118–23.

McIntyre, Matthew H., Emily S. Barrett, Rose McDermott, Dominic D. P. Johnson, Jonathan Cowden, and Stephen P. Rosen. 2007. "Finger Length Ratio (2D:4D) and Sex Differences in Aggression during a Simulated War Game." *Personality and Individual Differences* 42: 755–64.

Orbell, John, Tomonori Morikawa, Jason Hartwig, James Hanley, and Nicholas Allen. 2004. "'Machiavellian' Intelligence and the Evolution of Cooperative Dispositions." *American Political Science Review* 98(1): 1–16.

Ethical Challenges in Biological Research

Sex Differences in a Crisis Simulation Game

Rose McDermott

Investigation

Causes and sources of conflict preoccupy much of international relations theory. Oddly, very little of this research has focused on individuals. Most explanations find their origin in some aspect of the international environment or within particular state systems. And yet these institutions and organizations are comprised of individuals who, acting alone and in concert, often instigate or ameliorate situations of conflict and war. At this individual level of analysis, one of the most universal regularities across species and cultures lies in the fact that men engage in all kinds of physical conflict and fighting to a much greater degree than women (Daly and Wilson 1988; Flowers 1989; Wrangham and Petersen 1996).

In seeking to explore some of the reasons for why this may be the case, my collaborators and I have undertaken a series of experiments using simulated war games. As with many experimental research agendas, one set of findings inevitably leads to new questions and new studies designed to test and probe the lacuna left from the previous work. As attempts to distinguish the social and biological bases of aggression have continued, these experiments have become much more biological in nature than originally expected. And, as a result, unexpected and important ethical challenges have arisen as well. After briefly outlining the history of this work, this chapter concentrates on some of the specific ethical issues that arise when an investigator collects biological materials, particularly inherently identifying genetic material.

HISTORY OF THE EXPERIMENTS

The simulated war game that we have employed represents a kind of intellectual grandchild of work conducted by Morton Deutsch beginning in 1967. Deutsch was one of the earliest scholars to undertake experimental research in the area of international politics (Deutsch et al. 1967). In work that presages subsequent game theory in many ways, Deutsch and his colleagues were particularly interested in learning ways to encourage cooperation between individuals and states. This explicit concern with applied questions reflected an early defining characteristic of political psychology (Deutsch 1983); the often liberal bias inherent in the nature of these questions remained a recurrent cause of concern for some (Tetlock 1994). Nonetheless, several characteristics of Deutsch's early studies went on to heavily influence the nature and topics explored in future experiments by others. Specifically, Deutsch et al. (1967) investigated five different behavioral strategies in a two-person laboratory game using a confederate responder. Subsequent researchers did not commonly employ accomplices to conduct their experiments, especially once the non-deception norms favored by behavioral economists came to the fore. In addition, the use of confederates can prove both expensive and time consuming. However, the use of dyadic games became a quickly established mechanism by which to study individual decision making in conflictual situations.

As someone who had worked with Morton Deutsch, Cheryl Koopman expanded and elaborated on some of his early games to examine the ways in which male graduate students at Columbia University made simulated decisions about international relations in simulated contexts. This work found that instructing subjects to strive for superiority, as opposed to striving for parity, in their military balance with their opponents significantly increased the amount of money these men allocated to defense spending. In addition, these subjects proved more likely to allocate money toward defense if they were the recipients of hostile messages from their opponent (McDermott, Cowden, and Koopman 2002).

LATER EXPERIMENTS

In later experiments at Cornell, we examined sex differences in responses to these conflict scenarios for the first time (McDermott and Cowden 2001). For these experiments, subjects received course credit for their participation. In this first round of these experiments, subjects were assigned to same-sex male or female dyads. Each player role-played as the leader of a fictitious country in conflict with their neighbor, whose leader was played by the other subject, over some newly discovered oil resources on a nearby island. Over the course of four rounds, each subject was asked to make certain decisions. Each "leader" was given the exact same allocation of resources in each round and asked to decide how many weapons to buy and how much money to invest in industrial production. In addition, each subject was asked to take an action, which included doing nothing, attempting negotiation, making a threat, or going to war. They could also send a message to their adversary if they chose to do so.

In this study, we collected several personality indicators in order to examine the degree to which some of the behavior we observed might result from observable and systematic individual differences. As a result, in every round, subjects were asked to assess their own and their opponents' levels of aggression, hostility, competitiveness, and trustworthiness. In addition, before play began, we had subjects fill out several personality questionnaires, including the Social Dominance Orientation inventory, the Machiavellian and need-for-achievement scales, a thrill-seeking measure, and anger and fear indicators as well.

We had hypothesized that the female dyads would spend less money on munitions and prove less likely to go to war, and in fact this turned out to be the case. With one hundred subjects, 58 percent of whom were male, we found a significant positive relationship between weapons acquisition and the likelihood of going to war. Significant sex differences showed that women purchased fewer weapons and were less likely to go to war. Indeed, men proved approximately four times more likely to threaten their opponent than women. And, most dramatically, male dyads emerged as 390 percent more likely to go to war than the female pairs. Moreover, men judged themselves and their opponents to be more hostile, more aggressive, and more competitive than did women. Interestingly, no sex differences emerged in perceived trustworthiness.

Provocative differences emerged in the tone of the messages that subjects sent each other, as well. In the original conflict scenario that subjects read at the outset, fighting had broken out on an oil rig in disputed territory where several workers representing nationals from both sides were killed in an ambush. No male pair ever mentioned this instigating event. Further, no male pair ever expressed a concern about the native islanders on the disputed territory. However, female messages frequently demonstrated tremendous concern with tracking down the perpetrators of these murders and bringing them to justice. In addition, many women, in their negotiations over the disputed territory, made explicit claims concerning the importance of maintaining porous boundaries between regions so that family members could easily visit across areas, if desired.

HARVARD EXPERIMENTS

In a later, second rendition of the game conducted at Harvard, subjects played a computerized version of the game in same-sex and mixed-sex dyads. In this study, supported by funding from the Department of Defense, we offered subjects monetary incentives for their participation. In this version, several additional biological indicators were collected. We Xeroxed hands to examine the effect of finger length ratio on behavioral manifestations of aggression (McIntyre et al. 2007). In addition, we took several samples of saliva to assay hormonal levels of testosterone and cortisol (Johnson et al. 2006; McDermott et al. 2007). We also added several additional personality questionnaires, including a depression inventory, narcissism and self-esteem scales, and a stress test.

In this study, several important sex differences emerged. One of the most interesting concerned overconfidence. We told each subject that about two hundred people would be playing this game and asked each subject to guess, on a scale from 1 to 200, how well he or she expected to do. We asked this at the outset of the game, as well as after its conclusion, when people already knew whether or not they had won their particular game. By and large, men tended to overestimate their likelihood of success. Most men guessed at the beginning that they would score somewhere among the top ten people who played the game. Most women, by contrast, guessed that they would perform around the mean, that is, about 100. When men won their game, they generally increased their number, unless they already guessed the ceiling, that is, 1. When they lost, they might decrease their number a bit, but not as much as would seem objectively warranted. That is, men who lost might move their estimated score from 1 to 10. On the other hand, when women won, they rarely raised their number, but when they lost, they decreased it noticeably.

What accounts for this pronounced sex difference in confidence? Several anthropologists have posited that adaptive advantages led to selection pressures for positive illusions and overconfidence in the human evolutionary past (Tiger 1979; Trivers 2000). They suggest that positive illusions, or optimistic overconfidence, was triggered under conditions of threat and emerged more strongly in males than females because men were the ones with a long evolutionary history of fighting. In particular, overconfidence might confer advantages through self-fulfilling prophesies in health, creativity, and performance in the face of obstacles. Since historically, men have been the ones to fight in war, overconfidence may have proved adaptive because it helped motivate others to join in a coalition, increasing the probability of success by expanding numbers. In addition, such cognitive and motivated biases toward self-aggrandizement, illusion of control over events, and a pronounced sense of invulnerability to risk can increase combat performance by hardening resolve and increasing the ability to successfully bluff opponents (Johnson, Wrangham, and Rosen 2002; Trivers 2000).

For our purposes, we wanted to understand some of the biological mechanisms that might underlie positive illusions, and we suspected that testosterone might play a key role in mediating optimism in the face of battle because it can promote dominance responses in the face of challenges. Our results, while not definitive in this regard, remained suggestive. Pre-game levels of narcissism were related to self-ranking; in other words, more narcissistic individuals ranked themselves more highly prior to the game. Moreover, no relationship between self-ranking and actual performance emerged. It was not the case that those who thought well of themselves did so for good reason or because they were well calibrated with regard to the self-perception of their skills. Rather, the exact opposite proved true. Those who expected to do the best at the outset actually performed the worst.

As expected, testosterone was significantly related to pre-game self-ranking but only between, not within, sex groupings. In other words, men had higher self-ranking and higher testosterone as a group than women. But it was not the case that this relationship held up as statistically significant within each sex group, such that those men or women with higher testosterone were significantly more likely to rank themselves highly within each sex group. As a result, it is not possible to conclude that testosterone, as opposed to some other ineluctable aspect of being male, engendered narcissism and overconfidence in this population. Interestingly, with regard to the suppositions raised by the anthropologists mentioned above, levels of testosterone before the game were significantly negatively correlated with depression and stress and positively correlated with self-esteem. In other words, higher levels of testosterone made individuals feel less depressed, less stressed, and much better about themselves.

However, in this sample of 180 subjects, narcissism emerged as the only personality variable to affect the likelihood of a subject engaging in unprovoked attack over and above sex. That is, while many subjects defended themselves if attacked, fewer were likely to launch an unprovoked attack on their adversary. Being male significantly increased the statistical odds of engaging in this belligerent action. The only other factor that accomplished this goal as well was high narcissism.

Other personality factors demonstrated sex differences, however. Men tended to overrate their own, and their opponents', levels of hostility, competitiveness, skill, and intelligence more than women. Not surprisingly, assessments of one's own competencies in this regard proved more influential on behavior in both sexes than their assessments of their adversaries. Interestingly, ratings of self and opponent were statistically significantly related to initial testosterone levels, with higher testosterone leading to higher assessments of one's own hostility, competitiveness, skill, and intelligence.

Tone of communication in the messages that subjects exchanged exerted an independent effect this time around as well. Where hostile communications were present in dyadic interaction, both sides purchased significantly higher amounts of weapons, an average of twenty-one more battalions between the most friendly and most hostile pairs. In addition, hostile pairs allocated less money to industrial production, although in this version of the game, winning resulted directly from the amount of money in this account, which could be accumulated through negotiation or war. Finally, hostile dyads were seven times more likely to go to war than pairs that engaged in friendlier dialogue. Because this aspect of the study became quasi-experimental as a result of human interaction beyond experimental control, it is impossible to determine the direction of causality in the relationship between tone of communication and conflict, but the outcome remains suggestive.

In the most recent incarnation of these war-game experiments at the University of California, Santa Barbara, we are investigating the relationship between the so-called warrior gene and behavioral propensity toward aggression and impulsiveness (Caspi et al. 2002; Meyer-Lindenberg et al. 2006). This gene, MAO-a, represents a naturally occurring functional polymorphism present in approximately 35 percent of the male population. This genetic variant has shown a moderate association between traumatic early life events and increased risk for violence and anti-social behavior. Men with low MAO-a activity who have experienced a larger number of traumatic life events in their first fifteen years of life appear to be at increased risk for physical violence later in life. This risk seems particularly pronounced if the traumatic events happened between the time the boy was eleven and fifteen. Clearly, such gene-environment interactions prove challenging to investigate, but we hope to examine the relationship between genetic vulnerability, environmental risk, and subsequent propensity toward violence and impulsiveness.

ETHICAL CONCERNS

Several important ethical concerns emerged over the course of this research. This discussion begins with some concerns raised by the testosterone study, and additional ones resulting from the current MAO-a research. Finally, archiving issues are addressed.

Harvard Study

In the initial study at Cornell, which exclusively involved paper-and-pencil measures, few out-of-the-ordinary ethical concerns were raised. However, because of various aspects of the later study at Harvard, more problems came to light. In one case, we consciously decided not to collect data in order to retain privacy, confidentiality, and anonymity for subjects. This came about because an earlier pilot study had involved videotaped interactions between opponents in the middle of the game, designed to mimic a kind of "summit meeting" event. We had hoped to be able to code non-verbal manifestations of dominance and aggression in these conversations according to Paul Ekman's Facial Affect Coding Scheme (FACS). However, several of our subjects raised some concerns about how their data could be kept confidential if it was tied to material that could be tracked back to the videotapes, which would have their faces readily visible.

In addition, the human subjects review board asked us what we planned to do with the tapes when we were finished with our study. While the natural and normal answer was to destroy the tapes, we remained uncomfortable with destroying data, particularly if others wanted to replicate the work or question some aspect of our analysis. A real tension exists between protecting one's subjects and maximizing one's own protection against future challenges to the analysis, validity, or interpretation of the data. In the end, we decided not to make videotapes of subjects interacting and dropped the summit meeting aspect of the script. No work was published from the initial pilot videotapes, thus rendering moot issues relating to our own intellectual protection. As a result, the original tapes were subsequently destroyed.

In the main portion of the study, we nonetheless collected extensive material from subjects. Specifically, after reading and signing a standard informed consent form, subjects had their hands Xeroxed for the finger length ratio part of the study. Then they gave a saliva sample for the testosterone and cortisol assays. All subjects' information was both anonymous and confidential, but we needed to tie each subject's information together using the same number so that it could later be analyzed in concert. Using a strategy employed by the Veterans' Administration hospital system, we had subjects use the last four digits of their social security numbers as their subject identification number. This number proves useful because subjects will remember it over time. In case they participate in a follow-up part of the experiment, investigators can be certain to keep each individual's materials accurately connected together. These numbers are not enough for any given individual to be identified in a unique manner, but in small populations they remain unique enough to be distinctive.

Late in the first day of running subjects in the main study, a young African-American woman arrived, and we gave her the standard consent form. After reading it, she came up to me and said that she was very uncomfortable with this study. As she phrased it, she said, "You have my fingerprints on the Xerox, you have my DNA in the test tube, you have the last four digits of my social security number as my subject ID, and you are funded by the Department of Defense. Why should I believe you that you won't turn this information over to the government or be pressured to do so by them?" I can only imagine how much more concerned she would have been had the videotaping still been part of the experiment; in that case, we would have had her face on tape as well.

As she spoke, I realized for the first time what the experience must feel like to a subject concerned about his or her privacy rights. I knew that I had no intention of recording her identity or turning over any identifying information to anyone. But she had absolutely no reason to trust me. I honestly believed that she had probably had reasons and experiences from her life history that made her more sensitive to these issues than some of my other subjects, either because of her gender or her race. I was interested in keeping her in the study, but I did not want her to be uncomfortable. More importantly, I realized that it was important to me to demonstrate to her that I was an ethical investigator; for some reason, her expressed concerns made her opinion of me matter to me, perhaps more than it objectively should have. So I asked her if she might be willing to participate if I changed some aspects of her data. She asked what I planned to do. I said that I only wanted to know the length of her fingers, so

I would black out her fingerprints with a marker in front of her. I told her she could pick any four-digit number she wanted as her subject identification, as long as she used the same number for all of her data so that it could be tied together. I also asked her if she had any additional suggestions that might make her more comfortable. She thought about it and agreed. I blacked out the fingerprints in front of her and decided to have two people (myself and my head research assistant) measure and record her finger lengths right then so that I could then shred the Xerox in front of her, which I did. She picked four numbers for her identification, which I never knew and never asked. And she seemed quite comfortable with the outcome.

This interchange taught me at least two very important lessons. First—and this is a lesson I have had to learn many times doing experimental work in many ways—it is almost impossible to see an experiment fully from the point of view of the subject prior to the actual study. As much as experimenters try, we are often so influenced by and invested in our particular design that we forget how a truly naive subject will approach the demands with which they are confronted. Second, when a subject feels uncomfortable about any aspect of a study, an opportunity for real learning emerges. I could have told her that she did not need to participate in the study, paid her fee, and let her go. But for some reason, I realized in that moment that it was very important to me that she believe I was running an ethical and above-board experiment. I needed to do something to prove to her that I took her concerns, and her confidentiality, seriously. In so doing, we were able to find a strategy that made her comfortable, saved her data for my study, and also allowed us to reach a mutual understanding about the integrity of the study.

UCSB Study

In the most recent round[1] of the study, several additional ethical issues came to the fore, which I would like to address separately in this section.

Over the course of the earlier study, I had become increasingly uncomfortable with certain aspects of our Department of Defense funding. While the amount of money had really been a tremendous asset, it did raise suspicions among some subjects about who would actually have ultimate control over their data and information. For me, the most stressful aspect of this funding came from the strict time deadlines that were imposed by the federal government. Not all experimental studies go according to plan, and I was always worried that if something went wrong, I would be in real trouble, and I might have to pay back the money or have some other negative consequence.

When I was offered subsequent funding from the Department of Defense for a follow-on study, it came with certain conditions that seemed impossible for me to meet. First, they imposed a time deadline that seemed unrealistic to me. Second, they wanted me to investigate a particular gene that was not of primary interest to me. In particular, they proposed funding for a study to examine the effects of vasopressin (AVP) on aggression. While such work clearly deserves consideration, I felt that Ernst Fehr and his group in Zurich were conducting major research in this and related areas that were much more likely to generate results sooner than I could. He is an extremely skilled, experienced, and productive researcher, and he has a larger laboratory group. In addition, because the institutional review board requirements in Switzerland are quite different, he is able to administer certain substances to informed, willing subjects that would be impossible to do in the United States. This often makes his research much more efficient and conclusive. As a result, I decided that I could not do a good job with this project in a timely fashion, and I declined this opportunity.

I was lucky that two of my colleagues, a biologist/political scientist and a geneticist, were able to obtain independent funding for our project examining MAO-a through a group in Switzerland of which they are both fellows, the Branco-Weiss Society for Science, based in Zurich. This money has primarily been used to pay subjects and to pay for expenses related to the transport and analysis of the saliva samples. I had several subjects inquire about the source of funding during this study, and most did not express an objection. One subject remained convinced, despite being shown evidence to the contrary, that his data would become available to the Swiss government; this did not, however, stop him from participating in the experiment.

In this round of the experiment, we decided not to Xerox hands at the same time that we collected saliva. And we decided not to undertake any kind of videotape analysis, at least partly because of ethical concerns related to inherently identifying material.

Several additional ethical issues deserve mention. First, we never kept a master list of subjects with identifying information that would allow us, or anyone else, to connect individuals with the data they provide. This is particularly important because in the state of California, such information is subject to legal subpoena, and the university will not go to court to protect this data. While such legal investigation is certainly not common, one can never be sure when an immigration or criminal justice inquiry might impose on a study involving a large number of young men, including many foreign nationals. As a result, the only way an investigator can be sure that legal authorities can never get control of this data is to not collect it in the first place.

In our experiment, this constraint posed a particular problem because it had to be conducted in stages for logistical reasons. We began the study by collecting saliva and having subjects take two basic questionnaires on traumatic early life events and aggression. We needed to do this because the geneticist doing the analysis in Rome was moving to London and needed to complete the analysis before a certain date. We did not have the design for the later parts of the study finished in time. So we went ahead with the first part of the study and asked subjects who might be interested in participating in the second part of the study to write down their email addresses on a separate sheet of paper. Of course, many of these addresses were non-identifying, and this information could not be connected to individual subject identification numbers in any way. After we used these addresses to recruit subjects a second time, we destroyed the email lists. Subjects who participated in the first part of the experiment but did not leave an email address could not be contacted to participate in the second phase. These subjects were lost to the larger study and constitute experimental mortality, reducing the statistical power of any results we obtain. This is the cost of not obtaining any identifying information for any subject. However, by so doing, we ensure complete experimental anonymity and confidentiality for all subjects.

Second, there are some ethical and practical issues involved in shipping biological samples outside the United States for analysis. Because of the nature of our collaborative network and the source of our funding, it was best that our DNA analysis take place in Italy. Italy turned out to be one of the most difficult countries in the world in terms of ease and efficiency of customs regulations. Recent rules passed in the United States, including the Patriot Act, exacerbated the difficulty of shipping biological samples across national borders. Specifically, under the rules and regulations of the Transportation Service Administration, biological samples cannot be shipped on a commercial airliner unless it comes from a certified "known shipper." In the age of terrorism, and given the concern surrounding possibilities for bioterrorism, this is certainly understandable. However, being certified as a "known shipper" is not an easy or straightforward enterprise; it involves, among other things, at least two separate visits from federal authorities to investigate and certify safety and security. If someone has not been designated a "known shipper," then he or she must send biological samples on cargo air shipments. While many such flights exist, some of the most obvious do not fly everywhere a researcher may want to go. For example, Federal Express does not fly into Italy. Thus, it is imperative that researchers who wish to collaborate with others who work abroad secure all their transport needs prior to the collection of data, especially if samples are fragile in nature or need to be refrigerated. Certain companies can aid in facilitating this process for a considerable fee. Alternatives involve locating interested collaborators within the United States, although domestic shipment of biological samples across state borders often encounters similar impediments.

Note that ethical issues in this matter cut two ways. On the one hand, certain regulations may cause difficult impediments for researchers. But on the other, insufficient attention to protecting these samples opens the possibility that unauthorized individuals could obtain DNA samples of subjects who have put themselves in a researcher's hands for their own nefarious purposes, whatever those may be. It remains critical for researchers to safeguard the security of any biological samples that have the potential to be misused by others.

Finally, I would like to return to the issue of the destruction and reporting of data. In some cases, as with the analysis of saliva for hormones or DNA, the sample is literally destroyed or used up in the process of the testing. There is nothing left to be destroyed. But in other cases, some data, such as paper-and-pencil questionnaires, may remain. Even if these results are only reported in the aggregate in published studies, investigators need to decide how long to keep such data, and in what format. Should individual data be destroyed and only aggregate data archived? This may ensure greater confidentiality for participants, but many journals now require that entire data sets, albeit anonymous, be posted on the web to allow others to replicate the analysis. Videotapes remain a particularly difficult form of data because they include inherently identifying information. Yet many researchers may want to keep this data in case someone questions some aspect of their analysis or interpretation.

In addition, investigators need to consider what happens to their data if they should die unexpectedly. Procedures need to be put in place to ensure that such data are treated with respect or destroyed appropriately. Few researchers may want to plan for such a possibility, but securing the confidentiality and anonymity of subjects requires some thought to this consideration in advance of any unexpected disaster. Researchers who may be concerned for any reason that their data or some aspect of their data may become subject to outside legal or governmental scrutiny must pay particular attention to the way in which their data are archived and protected from outside intrusion or intervention.

Biological data and samples provide unique information, which can prove extremely useful for all kinds of research, not the least of which involves medical innovations and development. But with great promise also comes great responsibility. Whenever a researcher enters new territory, unexpected developments can take place. While medical researchers and biologists have undertaken such work for a long time, this area remains relatively new for political scientists and other social scientists. As with any novel approach, new ethical challenges can arise. In this area, one of the most pressing issues that may yet arise revolves around what we as researchers do with the information we uncover.

Medical research in particular lies replete with examples of individuals who study a particular area because they want to discover the gene responsible for a particular illness in order to find out if they themselves are likely to be vulnerable to Huntington's disease, breast cancer, Alzheimer's disease, or some other horrible ailment. And there are always examples of people who try out the MRI machine first, only to find that they themselves have some anomalous brain structure. For political scientists, the results are not likely to be so direct.

However, what if we do come to discover some genetic component for aggression? Does this mean that individuals will then be screened for this gene before being allowed to have certain jobs? Or screened by the military, for example, to place certain kinds of people in different kinds of jobs within the military? Would those with the gene who are convicted of violent acts, such as domestic violence, be sentenced to longer prison terms or be less likely to obtain parole? Should such consequences follow from genetic susceptibility when environmental forces play a key role as well? What do we do with the information we uncover? Perhaps nothing will be found that would hold profound societal implications. But we need to consider such implications as we begin to investigate some of the biological facets of social and political outcomes and behaviors.

NOTE

1. These experiments involve, in alphabetical order, Jonathan Cowden, Giovanni Frazetto, Dominic Johnson, and myself. Leda Cosmides and John Tooby provided helpful comments.

REFERENCES

Caspi, Avshalom, Joseph McClay, Terrie E. Moffitt, Jonathan Mill, Judy Martin, Ian W. Craig, Alan Taylor, and Richie Poulton. 2002. "Role of Genotype in the Cycle of Violence in Maltreated Children." *Science* 297(5582): 851–54.

Daly, Martin, and Margo Wilson. 1988. *Homicide.* New York: Aldine de Gruyter.

Deutsch, Morton. 1983. "What is Political Psychology?" *International Social Science Journal* 35: 221–36.

Deutsch, Morton, Yakov Epstein, Donnah Canavan, and Peter Gumpert. 1967. "Strategies of Inducing Cooperation: An Experimental Study." *Journal of Conflict Resolution* 11(3): 345–60.

Flowers, Ronald. 1989. *Demographics and Criminality: The Characteristics of Crime in America.* New York: Greenwood Press.

Johnson, Dominic D. P., Rose McDermott, Emily S. Barrett, Jonathan Cowden, Richard Wrangham, Matthew H. McIntyre, and Stephen Peter Rosen. 2006. "Overconfidence in Wargames: Experimental Evidence on Expectations, Aggression, Gender, and Testosterone." *Proceedings of the Royal Society of London B (Biological Science)* 273: 2513–20.

Johnson, Dominic, Richard Wrangham, and Stephen Rosen. 2002. "Is Military Incompetence Adaptive? An Empirical Test with Risk Taking Behavior in Modern Warfare." *Evolution and Human Behavior* 23: 245–64.

McDermott, Rose, and Jonathan Cowden. 2001. "The Effects of Uncertainty and Sex in a Simulated Crisis Game." *International Interactions* 27: 353–80.

McDermott, Rose, Jonathan Cowden, and Cheryl Koopman. 2002. "Framing, Uncertainty and Hostile Communications in a Crisis Experiment." *Political Psychology* 23(1): 133–49.

McDermott, Rose, Dominic Johnson, Jonathan Cowden, and Stephen Rosen. 2007. "Testosterone and Aggression in a Simulated Crisis Game." *The Annals of the American Academy of Political and Social Science* 614(1): 15–33.

McIntyre, Matthew, Emily Barrett, Rose McDermott, Dominic D. P. Johnson, Jonathan Cowden, and Stephen P. Rosen. 2007. "Finger Length Ratio (2D:4D) and Sex Differences in Aggression during a Simulated War Game. *Personality and Individual Differences* 42: 755–64.

Meyer-Lindenberg, Andreas, Joshua W. Buckholtz, Bhaskar Kolachana, Ahmad R. Hariri, Lukas Pezawas, Giuseppe Blasi, Ashley Wabnitz, Robyn Honea, Beth Verchinski, Joseph H. Callicott, Michael Egan, Venkata Mattay, and Daniel R. Weinberger. 2006. "Neural Mechanisms of Genetic Risk for Impulsivity and Violence in Humans." *Proceedings of the National Academy of Sciences* 103(16): 6260–74.

Tetlock, Philip. 1994. "Political Psychology or Politicized Psychology: Is the Road to Scientific Hell Paved with Good Intentions?" *Political Psychology* 15: 509–29.

Tiger, Lionel. 1979. *Optimism: The Biology of Hope.* New York: Simon & Schuster.

Trivers, Robert. 2000. "Elements of a Scientific Theory of Self-Deception." *Annals of the New York Academy of Sciences* 907: 114–31.

Wrangham, Richard, and Dale Peterson. 1996. *Demonic Males: Apes and the Origins of Human Violence.* Boston, MA: Houghton Mifflin Company.

Chapter 3

"The Aesthetics of Reason"

Exploring the Psychology of Virtue

Michael L. Spezio and Adam Martin

Conversation

Adam Martin (AM): A little bit of background, if we could—when we talk about the psychology of things like ethics, where did you get started with that?

Michael Spezio (MS): When I was in high school and even in junior high, I had a strong interest in the link between courage and compassion because my grandfather on my mother's side was a Holocaust rescuer. He was an ordained Reformed minister and member of the Confessing Church, and he used the wine cellar at the church and arranged it so that there was plenty of hiding space. I don't think he did an incredible amount of stuff—I mean, obviously he didn't do enough to have gotten himself killed because he actually survived the war and went on to live—but he was very beloved in his community, in part because of his public preaching against the Nazis. The people he rescued were mainly Catholics and Jews and some people of Polish descent as well. He was in Germany; he was outside Frankfurt in a little town called Niedergruendau. I still have family near there.

So that's how I first got into it. I was really close with my grandfather, and I kind of grew up on the stories of his courage and his compassion, his solidarity with the people he was serving and his opposition and my grandmother's opposition to National Socialism. My grandmother was one of the only nurses around, and so she did get a little bit of deferential treatment, even from the Gestapo. She had some liberty. One story is that my grandfather had this illegal radio where he would listen to the BBC to get the real news of the war and that was death if they discovered you. My uncle Klaus, who was a young child at the time, was bribed by some of the Gestapo with chocolate, asking, "Does your father listen to English-language or foreign-language broadcasts? Where's the radio?" and then they came looking for it, and they were going to take him away and shoot him, and my grandmother sort of stood them down and said, "That's ridiculous. We don't do that. Who are you going to believe? Me or this little kid?" She had helped heal some of their friends and colleagues, and there were regular German army stationed all around the property. So this is kind of where it started for me, and the first awareness dawned around the time I was ten.

AM: Did you know that you were going to do this immediately, once you got to college and graduate school?

MS: No. Originally I wanted to pursue science, because I really loved the beauty that science could bring. I went to grad school for biochemistry—that was in Ithaca at Cornell University—and did my PhD, but a year out from finishing my PhD in protein structure and function, I had an interview at UCSF for a very interesting position. It was a joint position between a guy who did X-ray crystallography structure-based drug design and a guy who worked on tropical diseases, which I really wanted to work on because I thought that this was a way I could actually give back using my skills.

I went there and I got the impression that the guy who did the tropical diseases was really serious about what he was doing. The guy who was doing the structure-based drug design, all he was interested in, I think, was getting stuff that would make a lot of money. That didn't really feel right for me. I had been for a long time, since I was about ten, thinking about going to seminary and doing ordained ministry. I went ahead and finished my PhD because it would have been a total waste not to; I had put in about four years at that point. So I finished up the following year, and I believe I defended on November 22nd and started seminary on November 29th. That was a whirlwind week!

I started at Pittsburgh Theological, but it was basically there that I determined that eventually I would go back to do some kind of science that focused on the human mind and its relation to human personhood and to human personhood in terms of agency and moral action. I really first wanted to do practical ministry for a few years, to get a close-up view of what that work really is about, and try to actually do it.

AM: Did you actually pastor a congregation for a while?

MS: I did; I did one year, but I could only do one year—you always have a dual-body problem! My wife is an engineer, but she wanted to go back to graduate school for environmental studies, environmental history. She was accepted at the University of Oregon; I couldn't find any position in ministry in Oregon. Finishing up at seminary, I couldn't find any pastoral position out there at all. I didn't really want to go just do a post-doc in biochemistry; I would have been able to do that, but it was not something that, coming out of seminary, called to me. What I decided to do was take an interim year—she deferred her enrollment, and I did an interim year. I was an interim associate at a Presbyterian church in Wisconsin. I worked for one year and still couldn't find anything. So my wife and I, right around New Year's, took a week-long retreat, just the two of us, very intentional, and the resolution was that I would apply to grad school at Oregon, but for neuroscience.

This would be cutting short the pastoral ministry in favor of just getting right back into the neuroscience but also doing as much working toward ordination. I was still in the ordination process. I had finished seminary, but I hadn't received any formal call, which you need to be ordained in the Presbyterian Church. So I went to Oregon for neuroscience and while I was there, I co-taught a science and religion course on human origins and moral agency.

So that was in 1999, and in June of 2000, we put on a two-day symposium. We had theologian Langdon Gilkey come out and give the keynote address. We had Mark Johnson, who is in philosophy at Oregon, and Helen Neville in neuroscience at Oregon. We had representatives from the Nez Perce reservation; we talked about Kennewick Man, about the origins controversy there. We had philosopher of evolution Michael Ruse out. It was a really dynamic time.

I was finally ordained on October 31, 1999, specifically to work and teach in religion and science. I received this grant and then the university essentially hired me and my colleague from the humanities program in classics to co-teach it. So that's it. From then on, my main work at Oregon was on audio-spatial attention, but I knew that's not what I wanted to stay in, and in the spring of 2000—maybe it was the fall of 2000—I took a seminar with Mark Johnson called the "Aesthetics of Reason." I love that title! I recently asked him if I could use that, and he's like, "What? I didn't even remember that I had used that title for my seminar." But I think it's a brilliant title. That was a great seminar, and that's when I was first deeply exposed to social neuroscience and the philosophy. And so basically, that seminar inspired me and I found my direction.

AM: What did he mean by the title? It sounds good!

MS: He meant that in a social domain, reason cannot operate adaptively without an affective and aesthetic component. You can't separate reason and emotion. There is no reason/emotion dichotomy. He was working with texts from Antonio Damasio and Ralph Adolphs, who subsequently became my mentor at Caltech. I first went to do some hard-core, single-unit monkey work at UC Davis. After two years, the head of the project decided he wanted to move in a very different direction, one that wasn't right for me, so then I moved with Ralph, who himself was just coming from Iowa, to Caltech. Since then, I've been very fortunately able to enter areas that are really interesting to me.

AM: Tell me a bit more about the concept behind what a "moral exemplar" is, and how it is that you've been able to test that.

MS: Well, you can operationally define a putative moral exemplar in the laboratory by saying that you have people who have become more than five or six standard deviations away from the mean. So we actually have a metric where we classify people by their behavior. It turns out that when we ran over two hundred people in an experimental public-goods game, just normal everyday adults coming from the LA area, about eighty-five to ninety of those people never, ever contributed to the public good. Only about roughly ten contributed every single time. Another ten contributed most of the time. So you had this one big peak with nobody, and then a small hump where people contributed only a few times, and then the other people were all the way out here at the extreme end of the distribution. So you basically have a blank spot in your bar graph. There's nobody in this—there's not a continuum; it's basically a discontinuity. So you circle these people and you operationally define them as exemplars. Now, that only gets you so far; that's a way to potentially select people. But then you have to say, "Well, are these people exemplary across a number of different conditions?"

AM: Beyond just donating to the hypothetical public good?

MS: Exactly. Do they represent, conceptually, the same kind of commitments to values as do real-world exemplars? And that's about where the Latent Semantic Analysis (LSA) comes in. We have some internal criteria for internal validity, but via the LSA, we can quantify our behavioral paradigm exemplars, the "BPEs," as we call them. We can quantify the semantic relationship between their self-understanding and the self-understanding of real-world exemplars. If we can quantify that, that gets us closer to external validity because nobody doubts, for example, that Holocaust rescuers and L'Arche caregivers—which is the group we're working on as well—nobody doubts that they are exemplary in some real sense. They may not be moral exemplars or virtuous exemplars in the true, full Aristotelian sense, but they are exemplary in a real sense. The question then is, and this is where we have to look at it neuroscientifically: are these people that we have in the lab—assuming they are also much closer semantically to real-world exemplars, so we have some external validity in that regard—are these people motivated by and assisted by the exact same neural process and neural networks as the people who are over here, who are not giving at all? Or are they motivated by and helped along in their choices by a neural network that is not just quantitatively but qualitatively different?

We have analytical methods by which we can ask those questions. Those are called machine learning or pattern classification methods that we are looking at and we have developed. We're looking at that both in terms of the neural activation during the public-goods paradigm as well as the neural activation during the rescuer paradigm. Think of it as multiple regression, where your independent variables are the brain areas and your dependent variables are the behaviors. The question is, in the multiple regression, if you take the best solution for the exemplar, do any of the values, the betas for the independent variables, go to zero in order to explain the non-exemplars, or the "Rands," Ayn Rand types? We just call them Rands, or non-exemplars. Here's the question: Does any other area pop up that is really not there for the exemplars?

AM: Is *there an area?*

MS: We don't know; we're still analyzing the data. That's how we approach the exemplar question scientifically. We also approach the exemplar question interdisciplinarily by looking at the brilliant work of Linda Zagzebski. She's at the University of Oklahoma. We're taking her careful work on virtuous exemplarity and virtue theory, and we're kind of rejecting the Aristotelian notion that being exemplary in one virtue entails all the virtues. One of the things that virtue theory holds is that there are people who are more or less resilient in their behavior, in their eusocial, virtuous behavior across conditions, and so that's one of the questions we're asking empirically.

We do not think that research in personality supports the notion that there is no such thing as character. That's a gross distortion of the literature that has been done by some philosophers. But in fact, usually personality accounts for between 30 percent and 50 percent of the variance. So individual character difference accounts for 30 percent to 50 percent of the variance. Then, of course, situation accounts for most of the rest. So the question that we have is, if we study these people over here on the right, are they the ones that are responsible for this residual accounting of the variance by individual factors rather than situational factors? Is it due to an averaging in which in any group there are only going to be 2 to 5 percent of people who are exemplary?

Take as an example the Milgram experiments; everyone emphasizes the fact that 65 percent of the people went up to the very end and shocked people to death; but if you look, about 20 to 30 percent of the people in those early experiments said, "To hell with you! We aren't going to do this to another human being!" Those are the people we should be studying, in my view. It's really trying to get a handle on what might be different about those people. This idea that Kristen [Monroe] has, which I like very much—do they *perceive* things differently? We could actually do eye-tracking, giving them complex moral situations in video format and asking, "Do they look at the same kinds of things?" and how long they're looking at people's faces, versus people who are not exemplary. Is their very perception different from the get-go? This is a question, I think, that's of interest.

AM: It seems like the psychology of moral behavior has really taken off within the last decade or so, whether it's this, whether it's positive psychology . . .

MS: Yes, but it's not exemplar-focused, so there's Darcia Narvaez, there's Kevin [Reimer], and a couple of others, but most of the psychology of moral action is dominated by a Kohlbergian framework that's quite dependent on rational structure and weighing of options rather than character-based and virtue-based.

AM: I talked to Kevin about this; he had some issues with the Kohlbergian paradigm, quite a bit.

MS: Yes and so do we all, because the Kohlbergian paradigm, and the prevailing model there, does not give enough due to the internalization, in terms of disposition and character, the unification of conceptual and emotional networks. In fact, what it gives a lot of due to is rational weighing of what is the best outcome, and so on.

AM: The trolley car, the Heinz dilemma . . . ?

MS: Exactly. What it's doing is basically taking all value and placing it outside. So, if it's consequentialist in nature, well the value is outside in terms of the outcome. If it's deontological in nature, then it's outside because of some lawgiver or some set of laws. Romanus Cessario, who's a wonderful moral theologian—his book just came out this year, a new revised, updated version—shows that in fact moral theology has gotten very far from what it really should be. It's been too concerned with externalization of the values, rather than virtue-based moral theology, which is all about actualizing the internalization of dispositions that give rise to virtuous outcomes and virtuous behaviors.

AM: So if they neglect this kind of stuff in theology, by definition they neglect it in psychology?

MS: Well, it's not by definition, but there's a parallel emphasis in psychology and neuroscience on the externalization of value, as well as in theology—you either have your consequentialists or you have your deontologists. It basically undermines the human capacity to control its disposition in ways that are purpose-driven, purpose-led—so both deontology and consequentialism always have emotion warring against reason. Virtue theory is the only approach that has these two linked together in an adaptive system. You don't have to suppress your emotion in order to be moral, to be virtuous. In fact, if you suppress your emotion, virtue theory would say, you can't be virtuous.

AM: And Damasio would bear that out?

MS: It's not just Damasio; it's the entirety of social neuroscience that's bearing this out. You can't make adaptive judgments without involving networks that are clearly underlying complex representation of our emotions.

AM: You talk about character. In moral philosophy, it actually seems a bit controversial because it presupposes agency in a certain regard. There's a large controversy, at least biologically, as to what extent we can say we have agency or free will or anything like that.

MS: Oh, yes. You have people like Martha Farah and Mike Gazzaniga and others who want to extend way beyond what the data have shown them. They want to just basically rule out all of human freedom. Well, it's completely self-contradicting to even argue for that. At this point, we don't know nearly enough in regard to the science to say one way or the other; we certainly don't have any evidence that the brain is identical to the mind. We don't have any evidence that would rule out some strong forms of emergence. And at the basic level, we really need to keep in mind that Dan Wegner, Martha Farah, Mike Gazzaniga, and others, when they talk about this stuff, they don't really fundamentally mean it because they say this, but then they're like, "Oh, we have to keep this illusion of freedom." They don't justify it in a way that makes any sense. The bottom line, for me, is that all of them engage in relationships with other humans, and they say they value those humans because of the intrinsic value of those humans and the joy in the relationship. But none of that is true if what they're arguing holds.

AM: Is the discussion of agency or free will even a controversy in your field? Does it affect what you do?

MS: No, definitely not. It's honestly not even a controversy because most practicing neuroscientists haven't even given it serious thought. They've just adopted the positivism and the reductionism of their schooling, and they have not really ever given it serious philosophical thought. But the science will come, and the science will say what it says. We shouldn't be afraid of whatever it says. But we should keep in mind that whatever the science says, it has said it in a defined experimental context that is quite different than most of what we encounter in the real world. We have to recognize and at least be humble in that regard; we can't over-extend from our data. This is happening left and right in neuroscience right now, especially when neuroscientists claim to find philosophical resolutions or physical resolutions to philosophical questions.

AM: It's interesting you mention that. I thought that what Benjamin Libet had to say, in that regard, was particularly interesting, insofar as it appears to be that there is a great deal of automaticity for actions we are about to take—the idea that what makes us human is restraint.

MS: Yes. But I don't think fundamentally that that's what makes us human. What makes us human, I think, is our capacity to nurture compassion and to be nurtured in compassion. And once we're nurtured in compassion, sure, then that stuff becomes automatic, but what's wrong with that? Automaticity, and sort of implicit action, is not a threat at all. In fact, that's what happens much of the time. But to go from there to say that we have no capacity to direct which implicit dispositions we foster in ourselves and that we're incapable of being directed by others who also choose to show compassion or destructive tendencies toward us, just seems to me to be a non-starter.

AM: More generally, in terms of studying not just moral exemplars but any of this philosophy of virtue, even positive psychology kind of stuff—love, forgiveness, gratitude, altruism—is this a viable movement within psychology that gets its due?

MS: Well, that's a good question. I think currently one would be safe to say that the main funding agencies, the National Science Foundation, the National Institute of Mental Health—they are interested in these areas, but frankly most of the near-term work, in the next five years, is going to have to come from foundations. Because this perspective is itself so new, we actually have our work cut out for us. We have a lot of, I think, compelling ideas and interesting preliminary data, but we have to rigorously, and I mean *rigorously,* test these hypotheses. We have to do that, I think, in the near-term, with foundation funding before we can actually go ahead and say, "Look, here we have a viable research program that has resulted in good, peer-reviewed publications; please give me grant money because this is really a viable alternative and really covers the bases and explains the aspects of moral activity in a way that the other way of approaching it can't."

AM: I get the impression that some people feel that this kind of research is tainted in that regard, by the foundations that fund it. The Templeton Foundation, for example. Some people outright refuse to take part in anything that has Templeton's name on it, because they think they have some theological agenda.

MS: Right. So, for full disclosure, I am a former member of the board of scientific advisors for the Templeton Foundation. There's a lot of misconception with regard to the Templeton Foundation, mainly because of the unfortunate residue of positivism that inhabits most scientific programs and training. There is an antipathy toward philosophical and religious (or I would say "theological") inquiry that's rigorous, that's scientifically well-informed. That doesn't seem to make sense to people who come up through our prevailing systems of science education, especially at the graduate level.

AM: The only experience that they have or want to have with this kind of stuff would be Intelligent Design as a foil?

MS: And that's one big misconception. The notion that the Templeton Foundation ever claimed that "Intelligent Design" was true is completely false. But there is that impression out there. These are people who, for example, take the Galileo affair and read into the Galileo affair a conflict between science and religion. It wasn't a conflict at all between science and religion; if you read Galileo's letter to the Grand Duchess Christina, he says that he himself is more pious and more religious than his theological opponents. It was a controversy and a conflict between different schools of biblical exegesis and theological interpretation. Galileo viewed it as a conflict between religion correctly understood and idiots who could not understand the way that the early church fathers thought. He had almost no respect for the contemporary crop of church leaders and theologians; he had tremendous respect for what he would call Holy Scripture and the Church Fathers.

AM: So then, as now, a lot of the "fights" between science and religion are more about politics and publicity than they are about anything substantial?

MS: Well, they are. I think to a large extent the supposed historical war between science and religion, of which Galileo is held up as a paradigmatic example, is a complete distortion of the actual occurrences, and it just results because of antipathy toward religion, and because of some who attack science in the name of religion. I have to say, backing up, that scientists have been on the receiving end of a lot of very harsh and unjustified and really destructive criticism. Poor Carl Sagan was absolutely vilified by many fundamentalist Christians, and it was horrible! So in that sense, I don't place the entire responsibility at the feet of the scientists; they have very good reasons for being suspicious. What they don't have a good reason for doing is allowing their negative experiences to form into a hardened bias against religion of all kinds. That's just fundamentalism of another stripe.

AM: Especially when there are religious people who have as much intention as they do for producing solid empirical research?

MS: That's right.

REFERENCES

Cessario, Romanus. 2008. *The Moral Virtues and Theological Ethics, Second Edition.* University of Notre Dame Press.

Damasio, Antonio. 1994. *Descartes' Error: Emotion, Reason, and the Human Brain.* New York: Grosset/Putnam.

Kohlberg, Lawrence. 1981. *Essays on Moral Development, Vol. I: The Philosophy of Moral Development.* San Francisco, CA: Harper & Row.

Libet, Benjamin. 1985. "Unconscious Cerebral Initiative and the Role of Conscious Will in Voluntary Action." *Behavioral and Brain Sciences* 8: 529–39.

Milgram, Stanley. 1963. "Behavioral Study of Obedience." *Journal of Abnormal and Social Psychology* 67: 371–78.

Virtuous Courage

New Methods for the Interdisciplinary Study of Virtue

Kevin S. Reimer, Michael L. Spezio,
Warren S. Brown, Gregory R. Peterson,
James Van Slyke, and Kristen Renwick Monroe

Investigation

Ordinary people sometimes engage in unusually courageous behavior. Our interest here is in presenting a new method for analyzing courage, defined as action on behalf of others, including those in one's community, in the face of a high probability of harm to oneself. Such behavior is worthy of scientific interest, particularly on psychological and neuroscientific grounds. In this article, we outline an interdisciplinary approach to the study of virtuous courage relevant to decision making. We begin with developments in moral psychology, documenting the turn to study of virtuous exemplars through schemas. We next extend our discussion to work in decision neuroscience, detailing the role of emotions and appraisals in schemas. We then suggest how exemplary and ordinary characteristics of courage may be addressed empirically through methods that consider the alignment of real-world actions and goals. We conclude by presenting findings from a pilot project designed to model these processes in a computational knowledge model known as Latent Semantic Analysis (LSA; Landauer et al. 2007). Our discussion notes possible uses of this approach for science.

VIRTUOUS EXEMPLARITY IN MORAL PSYCHOLOGY

The dominance of reason in moral theory has existed since Kant. Within moral psychology, the field has ignored concepts, such as courage, that did not fall into the orbit of moral rationalism, an approach pioneered by Jean Piaget and reaching its apex in the work of Lawrence Kohlberg (Kohlberg 1984). Kohlberg emphasized morality as justice reasoning elaborated through rules pertaining to harm and rights (Helwig and Turiel 2002; Kohlberg 1984; Nucci 2006; Turiel 2006). Beginning with studies at Harvard in the 1960s, Kohlberg outlined the development of justice reasoning through childhood, adolescence, and adulthood. The principal method of assessing change involved intractable moral dilemmas. Dilemma reasoning influenced other fields interested in the empirical study of moral behavior, including neuroscience (Greene et al. 2004; Greene et al. 2001; Marx et al. 2007).

Despite the utility of moral rationalism, questions mounted over whether dilemmas adequately captured the breadth of everyday functioning (Campbell and Christopher 1996; Narvaez et al. 2007; Walker 2004; Walker and Hennig 2004; Walker and Pitts 1998; Walker and Reimer 2005). One group took aim at the close resemblance between rationalist definitions of morality and liberal ethics typical of the Western academic elite (Haidt 2007; Haidt, Koller, and Dias 1993). This led to a wave of cross-cultural studies on the universality of justice reasoning. Heated controversy emerged over findings that communitarian priorities were more important than justice in the moral reasoning of poor Brazilians and Brahmans from India (Shweder, Much, and Mahapatra 1997; Turiel and Wainryb 1994). Additional concerns surfaced with the discovery that individuals with grave injuries to the prefrontal cortex of the brain were demonstrably amoral or immoral yet scored in the normal range on the Heinz dilemma (Damasio 2002).

Perhaps the most influential objections were exacted on the inability of the rationalist project to consistently link dilemma reasoning with everyday behaviors widely acknowledged to promote moral, altruistic, or compassionate outcomes. Two critiques emerged over this problem.

The first critique observed that Kohlberg's project marginalized the importance of self-identity in moral behavior (Blasi 1984, 1990, 1993). An improved approximation of moral functioning might include self and identity in persons widely recognized for outstanding moral commitments. Subsequent attention was directed toward study of nominated moral exemplars, particularly individuals known for extraordinary humanitarian commitments (Colby and Damon 1992; Monroe 2004). Exemplar interviews offered a complex picture of maturity as moral identity or commitment consistent with a sense of self to lines of action that promote or protect the welfare of others (Atkins, Hart, and Donnelly 2004; Hardy and Carlo 2005; Hart, Atkins, and Ford 1998). Allied projects focused on adolescent exemplars living in underserved urban neighborhoods—difficult environments presenting stark developmental contrasts for the growth of moral identity. A key study of exemplar youth from Camden, New Jersey, found that individuals demonstrating outstanding moral behavior scored no differently from everyday, matched comparators on Kohlberg's dilemmas (Hart and Fegley 1995). Moral identity for exemplar adolescents was characterized by clear goals, a sense of self-stability across different contexts, and influence of mentor figures (Matsuba and Walker 2004; Reimer 2003; Reimer and Wade-Stein 2004). These characteristics were formative in the development of coherent identities capable of effecting change.

The second critique argued that researcher notions of morality (e.g., justice, harm, or rights) might differ from participant experience, making it appropriate to begin with everyday conceptions of the moral domain (Flanagan 1991). The argument called for methods outlining a grassroots account of behavior based on common moral understanding. Along these lines, Walker and Pitts (1998) invited a large sample of adults to identify adjectives describing a moral person. Adjectives were subjected to Q-sort and clustering statistics. The study found that moral adjectives were distinct from those of highly religious and spiritual persons. Morality occupied a focused knowledge domain or prototype in thought. Moral adjectives were strikingly similar to virtues (e.g., honest, has integrity, loving), distributed as six themes including principled idealistic, dependable-loyal, has integrity, caring-trustworthy, fair, and confident. Recognizing that moral maturity may comprise more than one virtue prototype, a follow-up study derived adjectives for three different hypothetically moral individuals, including just, brave, and caring profiles (Walker and Hennig 2004). Further work confirmed that virtues account for more variance on moral outcomes than rationalist judgments alone, particularly with behaviors such as volunteer service (Walker and Frimer 2007).

Taken together, these critiques challenge the hegemony of dilemma reasoning in moral psychology. Virtuous exemplarity focuses empirical interest on actions undertaken in the midst of everyday pressures, responsibilities, and risks.

A recent study provides details for this agenda, considering nominated exemplars and virtue adjectives taken from Walker and Pitts (1998). Virtue influence on exemplar moral identities was studied through schemas (Reimer, Goudelock, and Walker 2009). Schemas are structured parcels of knowledge from memory situating the self in relation to others. Schemas can give rise to scripts or conceptual representations of action sequences associated with particular social situations. Schemas and scripts need not be completely available to deliberative thought in order to exert influence. These offer insight into how individuals understand the self through relationships (Hardy and Carlo 2005; Hart and Fegley 1995; Reimer and Wade-Stein 2004). In the study, schemas were constructed by asking exemplars self-referencing questions on actions, goals, pride, and distinctness. These narratives were then compared to virtue adjectives in LSA, a computational model capable of making similarity judgments between words and texts (Landauer, Foltz, and Laham 1998; Landauer et al. 2007). The objective was to find out whether constituents of moral identity in exemplar schemas (i.e., actions, goals, pride, distinctness) are organized differently than everyday individuals when compared with virtue adjectives. Study findings suggested that virtues are directly implicated in the development of mature moral identity. Exemplar schemas showed close proximity between actions and goals when compared with virtue adjectives in LSA. Schemas modeled in the same manner for comparators did not demonstrate this relationship.

THE NEUROSCIENCE OF MORAL ACTION

This psychological account of moral action is only a preliminary step. As shown, exemplar schemas represent a potentially valuable perspective on virtue and behavior. Yet schemas offer comparatively little detail regarding brain function that animates interchange between virtue, actions, and goals.

One recent proposal for emotions in decision making is known as Recurrent Multilevel Appraisal (RMA; Spezio and Adolphs 2007). Virtuous courage may function in ways that are significantly established within, but not limited to, emotions and appraisals. In offering a framework for decisions, RMA integrates thinking from two proposals in neuroscience: Appraisal Theory (Ellsworth and Scherer 2003; Lazarus 1991; Scherer 2003) and the Somatic Marker Hypothesis (Damasio 1994; 1999). We briefly outline each of these theories as background to our discussion of RMA.

A central concern for decision neuroscience pertains to the relationship between appraisal (perception) and emotion. Appraisal Theory (AT) is principally concerned with the manner by which individuals weigh appraisals in relation to personal goals, including the capacity to cope with consequences related to behavioral options available in a given situation (Scherer 2003). In this view, emotion plays a less central role in the decision process—appraisal is completed first, followed by emotion. The AT system is unidirectional, meaning that emotion does not exert reciprocal influence on goal-oriented appraisals. The Somatic Marker Hypothesis (SMH) establishes an alternative view of decision, with emotion taking the stage before appraisal (Damasio 1994). Emotion categorizes an event by constructing a link with known categories on the basis of prior experience. Appraisal therefore follows in the wake of reactivated emotion signals. Goals do not figure prominently in the process by which emotions are associated with these different experiences. The link between emotion and appraisal happens mostly beyond deliberative awareness, meaning that much of the decision process is attributed to bodily inference commonly understood as "gut" intuition (Damasio 1994; Haidt and Bjorklund 2008).

A potential shortcoming for these accounts relates to complexity in the relationship between emotions, appraisals, and goals. Both theories are premised on unidirectional influence. That is, neither accommodates the possibility that emotion and appraisal are reciprocally engaged toward selection of a contingency recognized as decision. RMA proposes a bi-directional association between emotion and appraisal. This is visible in two aspects. First, emotions become reciprocally engaged with cognitive elements of attention, such as memory retrieval, learning, and planning (Spezio and Adolphs 2007). Second, emotions contribute to appraisals, which also include personal goals and other goals residing outside deliberative reflection. Courage is established within emotion states reflecting appraisals laden with goal-oriented priorities. The source of these goals, according to the proposal here, is located in schemas. This proposed account suggests that courage, like other virtues, is not manifested in a singular emotion. Rather, courage is reliant upon dynamic exchange between different emotions and appraisals that align goals toward a particular course of action consistent with the self.

This discussion suggests that much of the decision process happens outside the direct influence of deliberative reflection. We are not, however, suggesting that deliberative reflection and attitudes are insignificant. Certainly these are important in framing situations at hand and providing an all-important "cognitive set" or perspective in which psychological and neural events take place. The current neuroscientific debate poses the question: is virtuous courage unconscious, deliberative, or both? A contemporary perspective on this question is found in Haidt's argument for morality premised on intuitions that function as evaluative feelings at the edge of consciousness. Moral intuitions are reflected in perceptions of social situations and behaviors, helping to regulate decisions on the basis of like-dislike or good-bad polarities. Disgust is centrally involved in orienting these intuitions. Haidt's Social Intuitionist Model (SIM) argues that moral reasoning is the product of a diverse range of intuitions catalogued in memory, providing us with emotional sensitivities that are subsequently recognized as decisions (Haidt and Bjorklund 2008). The intuitionist model assumes that everyday moral functioning is characterized by situations requiring rapid appraisals and responses beneath consciousness. For Haidt, the bulk of morality happens as an adaptive response to swiftly changing circumstances. Deliberative moral reasoning is perceived as maladaptive in this view, owing to its slow and sometimes painful evaluation of future contingencies. Such deliberation requires significant commitment of cognitive resources. From the SIM perspective, extensively reasoned decisions are typically experienced as a post hoc phenomenon, arising as a justification to intuition-driven decisions made rapidly and without extensive reflection. The SIM model links emotion (i.e., intuition) with unconscious decisions of moral importance. Appraisal is affiliated with conscious deliberation, albeit through explanations occurring after an event of moral significance.

Haidt assumes a unidirectional pathway of influence between emotion and appraisal. This exemplifies a classic dualism in neuroscience between "X"-systems of unconscious or automatic decision and "C"-systems of deliberative or controlled decision (Lieberman, Schreiber, and Ochsner 2003). While evidence exists for each system, it is probably better to regard these as mutually informing aspects of the decision process following the view of emotion and appraisal outlined in RMA. Unconscious mental processes are woven into deliberative decisions that are recursive.

Courage is deliberative in the sense that it is possible, via deliberative reflection, to frame subliminal emotional and goal-oriented processing. Framing that emphasizes deliberative, self-referencing engagement would, by the account offered here, lead to virtuous action.

A major scientific criticism of virtue-based moral action claims that phenomena such as courage reflect arbitrary, quaintly medieval assumptions divorced from real-world contingencies. The remaining section of the paper responds to this concern, examining the paradoxical nature of virtuous courage—comparatively rare (i.e., exemplary) behaviors that support the flourishing of ordinary people confronted by actual risk. With this paradox in view, we propose a new method for the study of virtuous courage. Pilot findings from the method offer an improved account of virtue through schemas in exemplars known for courage, namely individuals who rescued Jews during the Holocaust. The main goal is empirical validation of virtuous courage in schemas characterized by recursive interaction between emotions and appraisals.

VIRTUOUS COURAGE IN HOLOCAUST RESCUER SCHEMAS

Let us present pilot findings from a methodology that offers an account of schemas related to virtuous courage. Monroe (2004), in a well-known qualitative study of ethical action, documented the moral identities of Holocaust rescuers—individuals living in occupied Europe who risked their lives to shelter Jews from the Nazi concentration camps.[1] The narrative accounts of five rescuers document moral identity through agonizing, self-referencing decisions in the face of extraordinary personal risk. We hypothesize that Holocaust rescuers link actions with goals in schemas. We aim to outline the details of this association through unsupervised computational analysis of exemplar rescuer narrative relative to individuals who lived and fought in the European theatre of World War II.

Method

The pilot study was modeled on a matched comparator design common in the moral psychology literature (Colby and Damon, 1992; Hart and Fegley, 1995; Matsuba and Walker, 2004; Reimer and Wade-Stein, 2004; Walker and Frimer, 2007). In this design, nominated exemplars are matched on demographics to everyday individuals who lack exemplar behavioral criteria. The present study involved secondary analysis of existing data, namely, Holocaust rescuer transcripts from Monroe (2004). Holocaust rescuers were considered courageous exemplars from their actions on behalf of Jews in occupied Europe during World War II. These individuals risked imprisonment and death for their behavior. It proved a challenge to locate a suitable comparator group reflecting the unstructured narrative milieu of Holocaust rescuer data and rapidly shrinking pool of contemporaries in advanced age. The five Holocaust rescuers were matched on age, gender, and ethnicity to five English-speaking veterans or contemporaries of World War II as found in the archival narratives of an oral history series through National Public Radio.[2]

Procedure

For the pilot study, interview transcripts for five Holocaust rescuers were culled for courageous actions and personal goals. Recognizing the importance of schemas in the construction of moral identity, actions and goals were restated in the first-person singular. The process generated lists of actions and goals that were scrutinized for duplicate terms or words that might create artificially elevated similarity metrics in a computational semantic analysis. To illustrate, we attempted to avoid replication of words such as "care" or "assist" between actions and goals that might contaminate the analysis. Actions and goals that shared such terms were discarded. The final list included five courageous actions (e.g., "I made a hiding place") and five personal goals (e.g., "I wanted so badly to help"). This same procedure was replicated for the five oral history comparator transcripts. Comparator actions were not typically courageous in nature. No effort was made to exclude or include comparator actions based on apparently virtuous intentions. Retention of comparator actions was made solely on the basis of avoiding shared semantic content with goals. As before, the finalized comparator list included five actions (e.g., "I spent time in the local town") and five personal goals (e.g., "I want to be a strong man"). Actions and goals for both samples were subjected to computational semantic analysis.

The context for the methodology comes from work to specify the underpinnings of schemas, principally as representations reflecting semantic and episodic knowledge (Kihlstrom, Beer, and Klein 2003; Kihlstrom and Cantor 2000; Kihlstrom, Marchese-Foster, and Klein 1997). Semantic knowledge comprises abstract concepts regarding the self, including values, attitudes, traits, and motives (Kihlstrom, Marchese-Foster, and Klein 1997). Episodic knowledge includes concrete information regarding the self, such as events situated in autobiographical narrative. Both forms of knowledge are related to elements of memory (e.g., semantic and episodic) that support representations (Squire 1991; Tulving 1983). An early effort to model schematized representations was undertaken by Hart and Fegley (1995), who asked participants to rate their own narratives on bipolar personality traits. The process allowed researchers to map schemas for measurement of distances between representations. With its focus on narrative, the process preserved semantic and episodic knowledge for each participant. However, the rating system was laborious, requiring participants to make links between traits and narratives in a manner stretching the bounds of knowledge association. Even more problematic, participant ratings of descriptors missed simple associations in memory.

We sought to avoid these problems by comparing actions and goals with the LSA (Landauer et al. 2007) computational knowledge model. LSA makes unsupervised similarity judgments for any parcel of text, including narratives from Holocaust rescuers or comparators. LSA similarity judgments are mathematically obtained through a matrix decomposition technique related to factor analysis (http://lsa.colorado.edu; Landauer et al. 2007). Judgments are based on the model's global knowledge of the world, typically an 11-million-word first-year collegiate reader known as TASA. Based on this knowledge cache, the program assigns vectors to text parcels as approximations of meaning. Vectors are oriented on the presence of related words in TASA. As an example, the term "courage" most frequently co-occurs with "bravery," "endure," "battle," and "hero" in TASA. These neighbor terms provide meaning orientation for the "courage" vector. The model is capable of making fine distinctions in knowledge, such as the ability to grade undergraduate psychology essay exams with excellent reliability compared to human evaluators (Landauer et al. 2007).

Relevant to the abstract nature of virtuous courage in the pilot study, LSA demonstrates evidence of metaphor comprehension in everyday speech. The model "understands" metaphor by differentiating categories of non-literal and literal meaning. The model manages metaphorical meaning in natural language by making non-literal or literal categorizations in a manner similar to humans. Similarity measurement between actions and goals in LSA requires selection of a conceptually specific group of texts for comparison purposes. With the study hypothesis in view, proximity between semantic examples of actions and goals are meaningfully considered in participant schemas where the basis of comparison is virtuous courage.

One source of texts evincing courage is found in Walker and Hennig (2004), who derived a list of adjectives describing a prototypically brave or courageous individual. These courage prototype descriptors were adopted as the semantic basis for similarity comparison between actions and goals in schemas. A 60 × 10 LSA matrix was constructed for each sample group (i.e., Holocaust rescuer and comparator), respectively. LSA produced two covariance matrices, which permitted mapping of actions and goals reflecting schemas by sample group.

Results

To construct interpretive maps of actions and goals for each group, LSA covariance matrices were subjected to secondary multivariate statistics. Multidimensional Scaling (MDS) is appropriate for data where the basis for comparison between variables is undefined. MDS evaluates distance (similarity) between variables along orthogonal dimensions, resulting in a map or stimulus space (Hair et al. 2005; Kruskal and Wish 1978). Interpretation of variable relationships on the MDS map is potentially complicated by the reduction of multidimensional relationships into a two-dimensional plane. As a result, MDS is frequently paired with cluster analysis to specify relationships between variables. Hierarchical Cluster Analysis (HCA) provides a way of identifying relations between variables through clustering (Hair et al. 2005).

For the pilot study, MDS and HCA were together used to map distances between actions and goals where the semantic basis for similarity comparison was brave/courage prototype descriptors.

MDS. The ALSCAL MDS model was adopted for the present study as an approach suitable for data matrices of group samples (Hair et al. 2005). Generally, modeled data reflect good fit where Kruskal's stress value is ≤ .20 and R2 ≥ .60. For the Holocaust rescuer MDS solution, Kruskal's stress was .19 with R2 recorded at .75. These statistics suggested good model fit. For the comparator MDS solution, Kruskal's stress was .08 with R2 recorded at .98. These statistics suggested strong model fit. Thus, MDS generated two-dimensional maps indicating distances between actions and goals where these were compared with brave/courageous prototype descriptors in LSA. The MDS solution for Holocaust rescuers is given in Figure 3-1. The MDS solution for comparators is given in Figure 3-2.

Figure 3-1 Multidimensional Scaling Dimensions and Hierarchical Cluster-Analysis Clusters for Holocaust Rescuer Actions and Goals Compared with Courage Prototype Descriptors in LSA

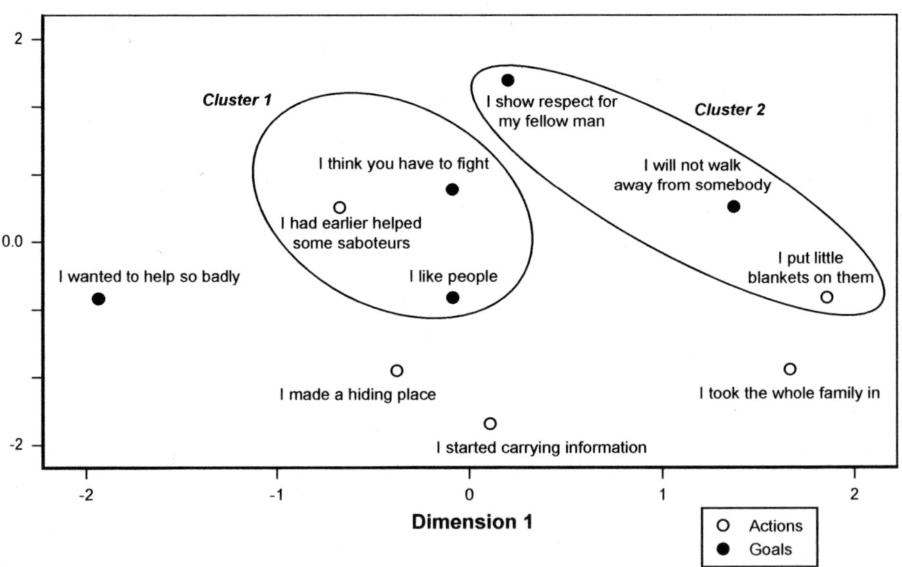

Figure 3-2 Multidimensional Scaling Dimensions and Hierarchical Cluster-Analysis Clusters for Comparator Actions and Goals Compared with Courage Prototype Descriptors in LSA

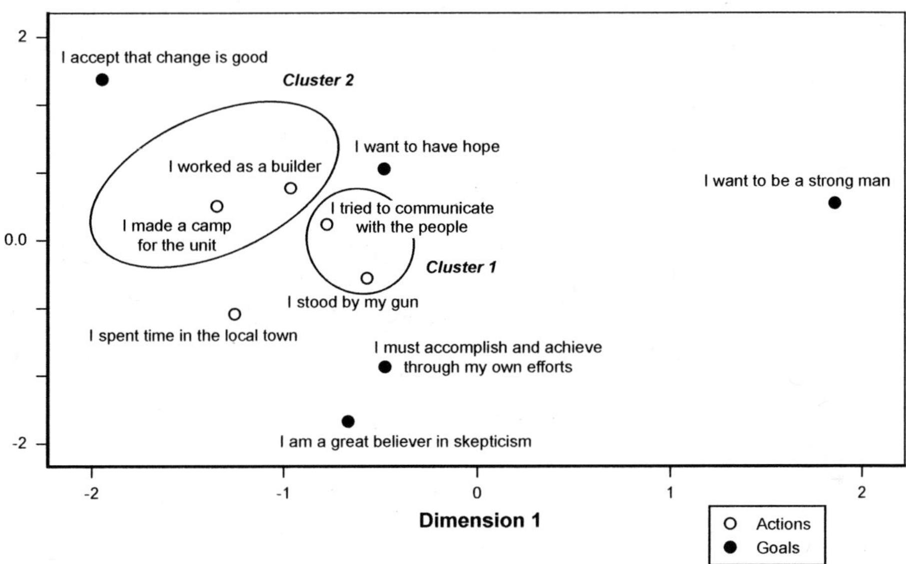

HCA. LSA covariance matrices were subjected to HCA, a multivariate technique that identifies cluster relations between variables. The pilot study used the most common HCA clustering procedure, known as the agglomerative method. This approach considers each variable as its own cluster. Subsequent steps combine closest variables. In some cases, a variable may become embedded within an existing cluster on a later step. Eventually all variables are grouped into a single large cluster. For the pilot study, an agglomerative HCA method was conducted using Ward's method with the squared Euclidean distance measure. It should be noted there is no established basis for identifying an optimal number of clusters for any solution, making the process somewhat open to interpretation. However, the HCA agglomeration schedule contains information that may be used to define clusters (Hair et al. 2005). The point where large percentage change between agglomeration coefficients becomes more stable (as reflected in smaller percentage change) is suggestive of how many clusters are indicated for the overall solution.

To simplify interpretation, HCA clusters are integrated into MDS maps in Figures 3-1 and 3-2. Using the percentage change in agglomeration rule, a two-cluster solution was indicated for both sample groups. Cluster boundaries are illustrated in MDS maps. Figure 3-1 presents clusters for Holocaust rescuers. Cluster 1 enclosed the action variable "I had earlier helped some saboteurs" with goal variables "I think you have to fight" and "I like people." Cluster 2 enclosed the action variable "I put little blankets on them" with goal variables "I show respect for my fellow man" and "I will not walk away from somebody." Figure 3-2 presents clusters for comparators. Cluster 1 enclosed the action variables "I stood by my gun" with "I tried to communicate with the people." Cluster 2 enclosed the action variables "I worked as a builder" with "I made a camp for the unit."

Discussion

The pilot study hypothesized that virtuous exemplars (i.e., Holocaust rescuers) integrate actions and goals in schemas. Close proximity between these cognitive representations presumably reflects recurrent multilevel appraisals that implicate emotion in the development of moral identity. This hypothesis was tentatively supported, recognizing significant limitations associated with pilot sample size and number of actions or goals subjected to computational semantic analysis. The LSA method shows promise as a technique capable of providing unsupervised assessment of narrative representations derived from people confronted with real-world contingencies requiring virtuous courage.

Our discussion is weighted toward issues associated with the novel conceptual and methodological framework, briefly considering (a) the relationship between actions and goals in schemas and (b) the need for a fuller methodological appropriation of emotion.

Actions and Goals. When evaluated within LSA, Holocaust rescuer actions appear to share brave/courageous semantic content with goals. These remarkable exemplars offer a version of courageous self-attribution that contrasts sharply with comparators. Exemplar actions and goals are powerfully other-oriented, sharing a common locus in schematized appraisal of social contingencies requiring a moral response. The actions and goals articulated by Holocaust rescuers are concrete, extending underlying values into relationships that bestow honor and respect upon others, thereby maximizing transparency of commitment to virtuous courage. The simplicity of a stated goal such as "I will not walk away from somebody" underlines the immediate, situational nature of schematized knowledge. Lofty moral principles may matter to these individuals in a general sense, but it is through embedded contexts of relationship that exemplars organize schemas that link actions and goals in a manner supportive of coherent moral identity.

Others have noted that chronically accessible knowledge is typical of mature individuals, potentially contributing to moral identity (Narvaez et al. 2007). Exemplar association between actions and goals implies that these individuals experience courageous attitudes and behaviors in a manner associated with outcomes impactful in the lives of others. Episodes of courageous character become entrenched in schemas, likely conferring a greater sense of Holocaust rescuer purpose and meaning. Meaning is a critical element in moral identity maturation, particularly when it contributes to legacy-promoting behaviors across the lifespan. A key facet of maturity visible through Holocaust rescuer schemas is their ability to realize personal goals of courageous effect. Successful manifestation of goals in action is reinforcing for this group, helping to explain why exemplars are able to persist in their commitments despite tremendous risks associated with their actions. Holocaust rescuers are realists who construct schemas through chronic knowledge of virtuous courage expressed through actions made meaningful through alignment with personal goals.

Emotion. A major shortcoming of the LSA methodology pertains to the inability of the model to appropriate emotion. While schematized organization of actions and goals presumably reflects emotional influences in the narrative experience of Holocaust rescuers and comparators, the details of emotional functioning remain oblique. The trick is to find an ethical proxy for virtuous courage that permits emotion assessment as a complement to LSA.

Modeled after the example of Holocaust rescuers, the rescuer game (RG) involves three different people (e.g., Perpetrator, Victim, Observer) playing ten rounds (Spezio et al. 2008). Prior to the first round, the Victim and Observer receive a separate allocation of money. On each round, the Perpetrator (an experimenter) takes a portion of the Victim's money. The Observer can see what the Perpetrator is doing and can decide to restore the stolen portion of money to the Victim. Over ten rounds, the likelihood that the Perpetrator will detect what the Observer is doing rises with (a) the number of times the Observer gives to the Victim and (b) the total amount of money the Observer gives to the Victim. If the Perpetrator detects the Observer's rescuing behavior, the Perpetrator will take all of the Observer's money. RG behavior potentially reflects emotional engagement with behaviors associated with virtuous courage. The contingency of observed victimization is combined with opportunity to intervene in a manner invoking actual risk for observer and victim. The RG represents a situation activating schemas affiliated with recurrent, multilevel appraisals.

Analogous to Holocaust rescuers, participants must consider whether action is congruent with personal goals or a distraction accompanied by emotions such as shame, guilt, or embarrassment. Real-world experiences and risks associated with RG anticipate decision framed by potential reward and punishment. Although the stakes in RG are reduced by many degrees relative to actual Holocaust rescuers, the Observer must nonetheless manage emotion-appraisal reciprocity affiliated with perceptions of injustice involving a living person and the possibility of monetary loss. Observer experience of the RG evokes powerful emotions that, following RMA, are dynamically engaged with appraisals involving actions and goals. Automatic and deliberative elements of decision are integrated into a mood reflecting the individual's fundamental commitments. The entirety of this process can be observed while RG participants play the game inside functional magnetic resonance imaging (fMRI) scanners, offering a means to considering emotion in virtuous courage.

NOTES

1. Rescuers risked torture and/or execution by sheltering Jews in their homes. See also Oliner (1992).
2. Transcripts were taken from the *This I Believe* series (http://www.thisibelieve.org).

REFERENCES

Atkins, Robert, Daniel Hart, and Thomas Donnelly. 2004. "Moral Identity Development and School Attachment." In *Moral Development, Self, and Identity,* edited by Daniel Lapsley and Darcia Narvaez, 65–82. Mahwah, NJ: Erlbaum.

Blasi, Augusto. 1984. "Moral Identity: Its Role in Moral Functioning." In *Morality, Moral Behavior, and Moral Development,* edited by William Kurtines and Jacob Gewirtz, 128–39. New York: Wiley.

———. 1990. "How Should Psychologists Define Morality? Or, the Negative Side Effects of Philosophy's Influence on Psychology." In *The Moral Domain: Essays in the Ongoing Discussion Between Philosophy and the Social Sciences,* edited by Thomas Wren, 38–70. Cambridge, MA: MIT Press.

———. 1993. "The Development of Identity: Some Implications for Moral Functioning." In *The Moral Self,* edited by Gil Noam and Thomas Wren, 99–122. Cambridge, MA: MIT Press.

Campbell, Robert, and John Christopher. 1996. "Moral Development Theory: A Critique of Its Kantian Presuppositions." *Developmental Review* 16: 1–47.

Colby, Anne, and William Damon. 1992. *Some Do Care: Contemporary Lives of Moral Commitment.* New York: Free Press.

Damasio, Antonio. 1994. *Descartes' Error.* New York: Putnam.

———. 1999. *The Feeling of What Happens: Body and Emotion in the Making of Consciousness.* New York: Harcourt Brace.

Damasio, Hanna. 2002. "Impairment of Interpersonal Social Behavior Caused by Acquired Brain Damage." In *Altruism and Altruistic Love: Science, Philosophy, and Religion in Practice,* edited by Stephen Post, Lynn Underwood, Jeffrey Schloss, and William Hurlbut, 264–71. New York: Oxford University Press.

Ellsworth, Phoebe, and Klaus Scherer. 2003. "Appraisal Processes in Emotion." In *Handbook of the Affective Sciences,* edited by Richard Davidson, Klaus Scherer, and H. Hill Goldsmith, 572–96. New York: Oxford University Press.

Flanagan, Owen. 1991. *Varieties of Moral Personality: Ethics and Psychological Realism.* Cambridge, MA: Harvard University Press.

Greene, Joshua, Leigh Nystrom, Andrew Engell, John Darley, and Jonathan Cohen. 2004. "The Neural Bases of Cognitive Conflict and Control in Judgment." *Neuron* 44: 389–400.

Greene, Joshua, R. Brian Sommerville, Leigh Nystrom, John Darley, and Jonathan Cohen. 2001. "An fMRI Investigation of Emotional Engagement in Moral Judgment." *Science* 293: 2105–08.

Haidt, Jonathan. 2007. "The New Synthesis in Moral Psychology." *Science* 316: 998–1002.

Haidt, Jonathan, and Frederick Bjorklund. 2008. "Social Intuitionists Answer Six Questions About Moral Psychology." In *Moral Psychology,* vol. 3, edited by Walter Sinnott-Armstrong, 181–217. Cambridge, MA: MIT Bradford.

Haidt, Jonathan, Silvia Koller, and Maria Dias. 1993. "Affect, Culture, and Morality, Or Is It Wrong to Eat Your Dog?" *Journal of Personality and Social Psychology* 65: 613–26.

Hair, Joseph F., Bill Black, Barry Babin, Rolph E. Anderson, and Ronald L. Tatham. 2005. *Multivariate Data Analysis,* 6th ed.. Saddle River, NJ: Prentice-Hall.

Hardy, Sam, and Gustavo Carlo. 2005. "Identity as a Source of Moral Motivation." *Human Development* 48: 232–56.

Hart, Daniel, Robert Atkins, and Debra Ford. 1998. "Urban America as a Context for the Development of Moral Identity in Adolescence." *Journal of Social Issues* 54: 513–30.

Hart, Daniel, and Suzanne Fegley. 1995. "Prosocial Behavior and Caring in Adolescence: Relations to Self-Reference and Social Judgment." *Child Development* 66: 1346–59.

Helwig, Charles, and Elliot Turiel. 2002. "Civil Liberties, Autonomy, and Democracy: Children's Perspectives." *International Journal of Law and Psychiatry* 25: 253–70.

Kihlstrom, John, Jennifer Beer, and Stanley Klein. 2003. "Self and Identity as Memory." In *Handbook of Self and Identity,* edited by Mark Leary and June Tangney, 68–90. New York: Guilford.

Kihlstrom, John, and Nancy Cantor. 2000. "Social Intelligence." In *Handbook of Intelligence,* edited by Robert Sternberg, 359–79. New York: Cambridge.

Kihlstrom, John, Lori Marchese-Foster, and Stanley Klein. 1997. "Situating the Self in Interpersonal Space." In *The Conceptual Self in Context: Culture, Experience, Self-Understanding,* edited by U. Neisser, 154–75. New York: Cambridge.

Kohlberg, Lawrence. 1984. *Essays on Moral Development: Vol. 2. The Psychology of Moral Development.* San Francisco: Harper and Row.

Kruskal, Joseph, and Myron Wish. 1978. *Multidimensional Scaling.* Newbury Park, CA: Sage.

Landauer, Thomas, Peter Foltz, and Darryl Laham. 1998. "Introduction to Latent Semantic Analysis." *Discourse Processes* 25: 259–84.

Landauer, Thomas, Danielle McNamara, Simon Dennis, and Walter Kintsch. 2007. *Handbook of Latent Semantic Analysis* (University of Colorado Institute of Cognitive Science). Mahwah, NJ: Erlbaum.

Lazarus, Richard. 1991. "Cognition and Motivation in Emotion." *American Psychologist* 46: 352–67.

Lieberman, Matthew, Darren Schreiber, and Kevin Ochsner. 2003. "Is Political Cognition Like Riding a Bicycle? How Cognitive Neuroscience Can Inform Research on Political Thinking." *Political Psychology* 24: 681–704.

Marx, Benjamin, R. Soames Job, Fiona White, and J. Clare Wilson. 2007. "Moral Comprehension and What It Might Tell Us about Moral Reasoning and Orientation." *Journal of Moral Education* 36: 199–219.

Matsuba, M. Kyle, and Lawrence J. Walker. 2004. "Extraordinary Moral Commitment: Young Adults Working for Social Organizations." *Journal of Personality* 72: 413–36.

Monroe, Kristen Renwick. 2004. *The Hand of Compassion: Portraits of Moral Choice during the Holocaust.* Princeton, NJ: Princeton University Press.

Narvaez, Darcia, Daniel Lapsley, Scott Hagele, and Benjamin Lasky. 2007. "Moral Chronicity and Social Information Processing." *Journal of Research in Personality* 40: 966–85.

Nucci, Lawrence. 2006. "Education for Moral Development." In *Handbook of Moral Development,* edited by Melanie Killen and Judith Smetana, 657–81. Mahwah, NJ: Erlbaum.

Oliner, Samuel. 1992. *The Altruistic Personality: Rescuers of Jews in Nazi Europe.* New York: Touchstone.

Reimer, Kevin. 2003. "Committed to Caring: Transformation in Adolescent Moral Identity." *Applied Developmental Science* 7: 129–37.

Reimer, Kevin, Brianne Goudelock, and Lawrence J. Walker. 2009. "Developing Conceptions of Moral Maturity: Traits and Identity in Adolescent Personality." *Journal of Positive Psychology* 4(5): 372–88.

Reimer, Kevin, and David Wade-Stein. 2004. "Moral Identity in Adolescence: Self and Other in Semantic Space." *Identity* 4: 229–49.

Scherer, Klaus. 2003. "Introduction: Cognitive Components of Emotion." In *Handbook of the Affective Sciences,* edited by Richard Davidson, Klaus Scherer, and H. Hill Goldsmith, 1–21. New York: Oxford University Press.

Shweder, Richard, Nancy Much, Manamohan Mahapatra, and Lawrence Park. 1997. "The Big Three of Morality (Autonomy, Community, Divinity) and the Big Three Explanations of Suffering." In *Morality and Health,* edited by Allan Brandt, 119–69. Florence, KY: Taylor & Francis/Routledge.

Spezio, Michael, and Ralph Adolphs. 2007. "Emotional Processing and Political Judgment: Toward Integrating Psychology and Decision Neuroscience." In *The Affect Effect: Dynamics of Emotion in Political Thinking and Behavior,* edited by W. Russell Neuman, George Marcus, Ann Crigler, and Michael MacKuen, 71–95. Chicago: University of Chicago Press.

Spezio, Michael, Warren Brown, Gregory Peterson, Kevin Reimer, and James Van Slyke. 2008. "Virtuous Decisions: Exemplarity in and out of the Laboratory." Poster presented at the Society for Neuroeconomics, Park City, UT.

Squire, Larry. 1991. "Memory and the Hippocampus: A Synthesis from Findings with Rats, Monkeys, and Humans." *Psychological Review* 99: 195–232.

Tulving, Endel. 1983. *Elements of Episodic Memory.* New York: Oxford.

Turiel, Elliot. 2006. "The Development of Morality." In *Handbook of Child Psychology: Vol. 3, Social, Emotional and Personality Development,* edited by Nancy Eisenberg, William Damon, and Richard Lerner, 789–857. Hoboken, NJ: Wiley.

Turiel, Elliot, and Cecilia Wainryb. 1994. "Social Reasoning and the Varieties of Social Experiences in Cultural Contexts." In *Advances in Child Development and Behavior,* vol. 25, edited by Hayne Reese, 289–326. San Diego: Academic Press.

Walker, Lawrence J. 2004. "Gus in the Gap: Bridging the Judgment-Action Gap in Moral Functioning." In *Moral Development, Self, and Identity,* edited by Daniel Lapsley and Darcia Narvaez, 1–20. Mahwah, NJ: Erlbaum.

Walker, Lawrence J., and Jeremy Frimer. 2007. "Moral Personality of Brave and Caring Exemplars." *Journal of Personality and Social Psychology* 93: 845–60.

Walker, Lawrence J., and Karl Hennig. 2004. "Differing Conceptions of Moral Exemplarity: Just, Brave, and Caring." *Journal of Personality and Social Psychology* 86: 629–47.

Walker, Lawrence J., and Russell Pitts. 1998. "Naturalistic Conceptions of Moral Maturity." *Developmental Psychology* 34: 403–19.

Walker, Lawrence J., and Kevin Reimer. 2005. "The Relationship Between Moral and Spiritual Development." In *The Handbook of Spiritual Development in Childhood and Adolescence,* edited by Peter Benson, Pamela King, Linda Wagener, and Eugene Roehlkepartain, 265–301. Newbury Park, CA: Sage.

Chapter 4

About Race and Politics

Jennifer L. Hochschild and Kristen Renwick Monroe

Conversation

Kristen Monroe (KM): How did you become interested in race and politics?

Jennifer Hochschild (JH): The historical answer, which I think is only part of the kind of answer you want, is that I wrote my dissertation book on American beliefs about distributive justice and poverty and inequality. I very carefully interviewed whites only; I selected out anybody who wasn't conventionally white. When the book was written people said, "It's a perfectly good book, but you can't write about equality, fairness, justice, and redistribution in the United States without thinking about race—they're so totally intertwined." I didn't want to think about race, which was why I interviewed white people, but people said, "Okay, but you can't get away with that." I therefore wrote the next book about race and started down this path. (Again, many things intervened and I could give a longer account.)

My underlying interest was, and actually remains, class issues. There's some funny irony in my research, which is that when I started, I was probably sort of a soft Marxist—I don't think I was a hardcore, committed, class revolution-type Marxist, but the core intuition, that *the* big divide that shapes people's life chances is class, has probably always been my conviction and in some sense still is. Now, why have I studied race for the last thirty years, given that my core conviction is about class? That's complicated. The intellectual answer is because class and race are so totally intertwined in American society that you can't really study one without the other (although I did for the first book). There must be some other more psychological explanation, which I don't quite have a handle on. Obviously I believe that race is crucial and shapes American history, government, politics, identity, social life, economic life, and anything in cultural life, but it's the intersection of race and class that I find most important, and that I really deeply care about. So I'm a little bit at odds with much of the race field for exactly that reason because I don't actually think that race per se is the biggest problem. I mean, the crude way of putting it is that you're better off being a middle-class black than a poor white; yes, you have trouble getting a cab in the middle of the night in New York, but at least you think about getting a cab, whereas if you live in Appalachia . . . ! Now, having said that, the 1995 book is about race *and* class, the intersection between the two. The logic of that book is to compare middle-class blacks, middle-class whites, poor blacks, and poor whites. The survey data in the book is divided into pretty straightforward two-by-two tables, and almost all of the empirical analysis is either that sort of two-by-two table or some particular comparison within it. So that book, which I think is probably the best thing I've written, is explicitly about the intersection between race and class, and maybe it's the best thing I've written because that's where my real core conviction lies—or maybe because I wrote it when I was forty rather than twenty or sixty, who knows.

KM: So are you going to go back and look more at class?

JH: Well, that's what the current book manuscript begins to do. The working title, which will surely change when the marketing department gets hold of it, is something like, *Transforming the American Racial Order*. And the core intuition is that we are at a volatile moment in American racial politics—Obama being both symbol and cause, but it's a much broader claim. The central argument is that there are a set of powerful forces that put us in a position to be able to transform the American racial order. (We have a definition of the racial order that has five components, used to structure the argument through the book.) There are four key driving forces, the first being immigration, which brings a new population to the country, changes the way we understand race, changes the options for a whole bunch of people, changes the demographic and therefore eventually the political structure, perhaps. The second force is multiracialism as a self-defined identity, the deliberate choice not to be in one box or the other but to identify as mixed race, which implies a degree of flexibility and fluidity and a whole different understanding of what it means to be "raced."

The third component is genomics, which I think is going to be the driving scientific dynamic of the twenty-first century. On the one hand, it has the potential to "re-racialize" a whole lot of things, to the degree that there are claims of a biological component to what we think of as conventional races. On the other hand, the optimistic story is that in fact race will become largely irrelevant in a lot of social arenas—obviously medicine, maybe law, maybe others—because medicine, for example, will become entirely individualized. The fantasy, which might come true, is that ten or fifteen years from now, your genomic profile will be stapled to your birth certificate; it will cost $200 to get your whole genomic profile and your doctor will see exactly what's wrong with you, at least biologically, from your profile and give you the right medicine. So race will in effect disappear from the medical arena, except when people want to use it. And race could disappear from the legal arena—you don't have to have a guy in court; you have the blood sample from the crime scene, match it up with the FBI CODIS, and you know he did or didn't do it. Again, this is much too simplistic, since a sample from a crime scene could be there for a lot of reasons. But the main point is that genomics could either re-biologize race, or it could in fact transform the way we understand individuals, effectively to leapfrog right over it. So that's the third component.

The fourth component is cohort change, which really drives the other three in some ways. The core story is that eighteen- to twenty-nine-year-olds live in a different world than you and I. We all imprinted on the 1960s and so did people up through the early 2000s. However, when my son was eighteen, I asked him, "What's the civil rights movement?" He says, "I study it in AP US history." It didn't shape his life, as it did mine. We have a fair amount of survey data and some behavioral data showing that this generation is just different—they live in a different racial world from the one of older adults. Demographically and politically, their critical formative moment is going to be some combination of 9/11, the 2005 Superdome in Hurricane Katrina, and the election of a black man to the presidency. Now, who knows? Obama may be a fiasco; there's obviously a backlash, his political legacy is not simple—but his election will remain a defining moment in American racial politics, especially for young adults.

Anyway, we say that these four things in conjunction—immigration, multiracials, genomics, and cohort change—put us in a position to be able to transform the American racial order.

Now, having said that, there are blockages and we aren't *predicting* transformation. And that's where class comes in. We basically see four potential blockages. The first is highly disproportionate racially based incarceration, which could create a whole new racial-group hierarchy. That's very class inflected; it's very much poor, young black men disproportionately, and their families and their communities, who are just screwed. The second is wealth holding, which is by definition a class story, and which has a whole bunch of associations with jobs, with health, with life expectancy, with the opportunity to go to college—wealth holding is the key component with huge consequences for life chances. And that's clearly a class story. The third blockage is the possible creation of a new underclass, illegal immigrants being the most obvious case and possibly Muslims. In that scenario, we basically retain the same hierarchy and just slot some new people in the bottom, so blacks are a little better off because they're not illegal immigrants—that's a class story also.

KM: Do you think that's going to happen, or will blacks continue to be at the bottom of the heap?

JH: *Poor* blacks will. My conclusion to the book—I have two co-authors who don't necessarily agree with me—is that what we end up with is basically a class division, that middle-class blacks (and Hispanics) are mostly doing fine, still with issues of discrimination, but their problems are now relatively marginal, and I think there has been a genuine transformation. By middle-class I mean certainly the top 50 percent, maybe the top 60 or 70 percent. Not that discrimination and racism have disappeared, but their life chances are basically okay; it's a hassle, not a disaster. The bottom 10 to 25 percent—and you can debate the exact numbers—their lives are a God-awful mess. They're in jail, they're illegal immigrants themselves or live among them, they face good old-fashioned overt discrimination, their schooling is terrible, their jobs are insecure or nonexistent—their lives are very difficult. So we have transformed and will continue to transform the racial order in a very, very big—historically unprecedented—way, for maybe 75 percent of the population: almost all Asians, a very large proportion of Hispanics, half to two-thirds of blacks, most whites. And there will be a residual category at the bottom whose lives are just shitty, and they will be disproportionately black and illegal immigrants.

So that's the argument of the book. I'm reverting, in a slightly different way, to the race-class intersection, essentially by saying that what has changed is that most of the black population has, in effect, escaped from the worst elements of the traditional racial hierarchy. But it comes down like an absolute ton of bricks on the people who are left out.

KM: So you think that's what's going to happen?

JH: That's my guess, yes. And I think it gets worse and worse for the bottom *as* it gets better for the top. The bottom 10 percent get worse off as society moves from 50 to 60 to 80 to 90 percent who are better off, because they're left out. They're in jail; they're single parents; they're illegal. They all live in ten blocks in ten cities; it's extremely concentrated, and the more that middle-class blacks move into the suburbs and go to Harvard and UC Irvine and so on, the less they will feel—and be—connected with the blacks at the bottom of the economic heap. To the degree that racial solidarity has been a protective force for the population as a whole, it is breaking down. Racial solidarity is breaking down and class movement is taking over. So it's a new way of making the race-class argument, a little different from the old book, but it's the same preoccupation.

That's the optimistic story—the story I've just told you is the best that we can hope for and would be a very big change. I don't know if that's going to happen. The book ends up saying that it's really complicated and who knows, that it's hard to tell what's going to happen, that politics matters and we need to see what the next generation does—which is why it's going to be an academic press book rather than a popular press book, because we can't bring ourselves to say, yes, this *is* going to happen. And Vesla Weaver and Traci Burch, who are my co-authors, are not even willing to go as far as I just did in this description; they're still persuaded by the evidence that race, independent of class, is a more constraining phenomenon than I am willing to say. I'm white and I'm more optimistic; they're black and they're less optimistic—in our own little three-way authorship, we're playing out the issue that's out there in society.

KM: Is it also hard to be a white person in that field?

JH: It's salutary, partly because my department at Harvard, the African and African American Studies department, is really fine on this. I would say a third to half of members of the Harvard department are not black; some are mixed race, but there's probably around twelve of us in the department who are white, broadly defined. I learn a great deal from my colleagues, not always when they intend or realize it.

KM: What does "being a soft-core Marxist" mean to you?

JH: It means to me that class is the crucial social driving force, not gender or race.

KM: And do you believe there's a dialectical inevitability to this?

JH: No, and that's why I say soft-core. I mean, Marxist is the wrong term—"class-ist" is better, but we don't have a vocabulary to talk about what I think. Materialist? As in materialist rather than idealist in the old Hegelian sense? Maybe that's the right term. But no, I certainly don't think there's an inherent inevitability about it. We don't have a category for what I am.

KM: You're talking about people who are gut-wrenching poor, I think. You're saying that there's an underclass, a group of people who are trapped in a culture of poverty.

JH: Yes, I mean that, although I'm not sure how culture and structural possibilities intersect. I certainly am concerned about the people who are the bottom 10, 15, 20 percent—I have a lot of flexibility in what I count as the bottom. About the rest of the population, setting aside the devastatingly poor, I believe two contradictory things and I don't know how to reconcile them. I do believe—and I've been looking at the evidence for the last six months—that people largely end up where they start. So there *is* a class story in the broader sense, which is to say that the best guess of where somebody's going to end up is where they or their parents start. That's truer for blacks than it is for whites—whites are more likely to be upwardly mobile from where they start than blacks are—so there's a race-class intersection. Then the other thing I believe, which is contradictory, is that, if you look at the 1960s or the 1990s and the 2000s, many more blacks are in the middle class, or are successful by conventional standards, than they or their parents were in the past. How to reconcile those two things, I don't know. It's partly a generational story.

KM: Do you think that those gains will be permanent?

JH: Yes. I mean, nothing's permanent, but relatively so. This is all in terms of income, not wealth. Actually, the wealth data show much less change than the income data, or education, professionalism, lifestyle, home-ownership, or anything. Why those income-type measures haven't translated into wealth measures is a very big question to which nobody has a complete answer. What we also don't yet have is good data on how much disproportionately poorer blacks are as a consequence of the 2008 crash and the home ownership troubles. Most research shows that blacks were hit harder than whites. But the Federal Reserve Board of New York City says the opposite—they have a study of 75,000 homeowners and they say there's no evidence of disproportionate loss on the part of blacks, and if anything, there's a small amount of evidence in the other direction. I'm not a good enough economist to understand why their results are different from those of forty-seven or so other people. But I know the guy who did it and he is a fine economist; it's a Fed report, and it has to be taken seriously. Whether it outweighs all the other stuff, I don't know—nobody really knows, probably because the data are too new. But there's some plausibility to the Fed argument, in part because whites, on balance, owned houses that were worth a whole lot more, and so the loss of a house for the median white is a bigger chunk of their wealth holding than the loss of a house for the median black, who didn't have much wealth to begin with and therefore had a cheaper house. Or, maybe they didn't get the easy loans that the whites got, so they're less underwater, or they had less of an investment because they bought the houses much more recently, so they were only six months into a mortgage rather than fifteen years.

KM: And most of the big housing bubble was not in the inner-city.

JH: That's right. So it may turn out that the 2008 crash, in some terribly ironic way, actually improves black wealth-holding in relation to that of whites, in that both of them lost, but whites lost even more.

KM: I think you're someone who really cares about inequality and poverty, rather than class per se.

JH: I don't know what it means to say class *per se* rather than inequality and poverty. I mean, I'm certainly not a Marxist in any kind of robust sense, to think that there are structural inevitabilities, that there's a proletariat and a bourgeoisie. To the degree that it's about class, the evidence does suggest that people in general are more likely to end up where they started than to move up the class ladder in a substantial way, and blacks even more so than whites. That's something that looks more like a class story than a story about only poverty.

KM: And when you say "started," you're talking about economically?

JH: Yes. But as I said earlier, over the course of generations, that's changed, and the black middle class now exists in a way that it didn't in the 1960s. So my intergenerational story does not match my lifetime story. That's where the contradiction lies, and I don't quite understand how to reconcile that. If people begin and end more or less in the same place, but somehow thirty years later their kids are much better off, something has changed.

KM: Structural changes?

JH: Yes, that's part of it. More people in general now go to college or have white-collar jobs. So I think you're right that my deepest concern is with the bottom 10 to 20 percent. I think that's at least as much structural as it is cultural, probably more so—although I'm not prepared to say that there's no cultural, behavioral, personal, or individual component to it, which seems a silly argument to me.

I'm also concerned about—maybe the way to put it is in terms of wealth holding rather than class because wealth holding is what drives the chances for genuine upward mobility that's not structural. If you lose a job, do you lose your house? If your kid can get into a good college, can you afford to send him or her there? Can you take your kid to a museum if you feel like it? If your kid wants a book, do you have to wait until the library has a copy? You could give a zillion examples of how wealth matters.

KM: It sounds like you're talking about Amartya Sen's concept of capability, the idea that we should be talking about wealth not in terms of money but in terms of how capable a person is of living a long life, having access to education, good health, and so on?

JH: Yeah, it's pretty close to that. I think that's about right. But also, I'm never certain exactly how much race per se matters. Unlike many in this field, I'm not prepared, for example, to say that the Tea Party is racist, though I'm certainly prepared to say that there are racists in it. I'm certainly prepared to say that very conservative, middle-aged white Republicans are racially much more conservative than the rest of the American society. Are they racist? What does "racist" mean? That they don't like black people? Well, black people don't like Jews, Koreans don't like the Japanese—that's the world. And this is where, as I said, being white in an African American Studies department is salutary—many of my colleagues there and certainly other people in the political science profession reach for racism as an explanation for all kinds of outcomes that I am no longer prepared to say are *simply* about race—not that race is irrelevant; race is never irrelevant.

Here's a more academic way of making that point: it turns out that when you actually ask people about their racial linked fate, and then also about their class, or gender, or religion linked fate—which few have ever done before—class jumps out as important, much more than race. That's less true for blacks, however—back around the circle.

KM: I'm someone who thinks about innate dispositions and hates the whole concept of race. We're all members of the human race and skin color varies greatly according to all kinds of factors. One of the things that looks interesting to me about the United States is that part of what's happening is that there's more intermingling of the so-called races, and there are so many people who want to be called mixed race or whatever, and class doesn't seem to be that kind of concept—it's a much more economic concept.

JH: Yes, I agree with that. Here is another way in which class matters, in affirmative action: if you look at universities or law firms, we have a lot of representation of gender issues—as we should—partly because there are a lot of women there to articulate gender issues. We now have a lot of representation of race and ethnicity issues, but we don't have representation of class issues.

KM: If you're really poor, you're not even going to get to go to college.

JH: Right. And that's my frustration; we're fighting the old wars. And that's why this cohort change stuff is so interesting. Because it feels to me that there's some evidence that many twenty-five-year-olds are looking at the world differently. We'll see.

NOTE

This conversation was written with the assistance of Peter Hawkins.

REFERENCES

Hochschild, Jennifer. 1981. *What's Fair? American Beliefs about Distributive Justice.* Cambridge, MA: Harvard University Press.

———. 1995. *Facing up to the American Dream: Race, Class, and the Soul of the Nation.* Princeton, NJ: Princeton University Press.

Individuals versus Group?

The Moral Conundrum of Blurred Racial Boundaries

Jennifer L. Hochschild

Investigation

It should not be the government's responsibility to dictate to an individual how ... to identify themselves.... It is imperative that he [my child] ... has the freedom to identify as a multiracial person. This can only be accomplished by including a multiracial category on all forms that require this information.

—A mother testifying at the 1997 Congressional hearing on categories for US Census

A multiracial classification might disaggregate the apparent numbers of members of discrete minority groups, diluting benefits to which they are entitled as a protective class under civil rights laws and under the Constitution. In our quest for self-identification, we must take care not to recreate, reinforce, or even expand the caste system we are all trying so hard to overcome.

—Harold McDougall, spokesperson for the NAACP, 1997 Congressional hearing

What is right or good for an individual is not always right or good for his or her group. This conundrum is well known; Shakespeare, for example, uses it to ratchet up the tension in *The Merchant of Venice* when Jessica abandons her Jewish community to marry the Christian, Lorenzo. There will never be consensus on whether a person should follow self-chosen purposes or group-based commitments if the two values conflict. Nor will I try to resolve that dilemma; my goal in this chapter is to explore its ramifications in the arenas of racial identity and boundary-blurring.[1]

Of the many possible formulations of the broad conundrum, I focus on the particular version best identified as liberal individualism versus group-based communitarianism. By liberal individualism, I mean the political philosophy that stresses the primacy of individual freedom and rights, protected in the final instance by the state and its laws but more immediately by social norms that valorize autonomy and self-definition. By communitarianism, I mean the political philosophy that privileges the well-being of the community or group to which individuals adhere or are born into. It, too, might ultimately be backed by the state and its policies, but my primary focus here is on social norms and practices.

The two values need not conflict. Individuals may choose to place priority on or define themselves in terms of the good of their group; groups may be committed to fostering the development of each member. But the values frequently do conflict in the political realm, when people face a structure of choices in which group gains will come at the expense of individuals in the group, or individuals can attain their goals only at the expense of the group. This conflict may be especially common in the arena of racial boundary construction and dissolution. From the perspective of a bounded and disadvantaged group, such as African Americans or Francophone Québécois, boundary-blurring threatens exit in Albert Hirschman's classic trilogy of exit, voice, and loyalty (Hirschman 1970). The person who rejects strong group identification in favor of self-definition may be perceived as exiting—being disloyal and thereby diminishing the potential voice and political power of the group. If enough people exit, the whole group suffers and risks even deeper disadvantage. That is, if group members with the highest social status—such as light-skinned blacks or bi-national Canadians—move away from the group, it becomes easier for hostile outsiders to stereotype or discriminate against group members who remain within the (now smaller) group's boundaries. Thus in this view, exit or even intentional straddling (calling oneself a "mixed-race black"; McBride 1996) may be intended to break down group boundaries but actually helps to strengthen racial barriers. More analytically, movement across racial boundaries may shift the location of the boundary, but it strengthens rather than dissolving the boundary itself.

Conversely, from the perspective of an individual who seeks some distance from group identification or who wants to choose his or her own identity, group loyalty and voice may come at too high a cost. That person might prefer exit for private reasons—a marital choice; commitment to another, nonracial, group; higher social status; identification with a parent, ancestor, or culture outside the group. Or that person might believe that the best way to abolish racial hierarchy is to dissolve race-based groups in favor of all individuals' freedom to define themselves as they wish. In this view, group loyalty and voice may, or even must, give way to the right of exit.

In this chapter, I explore this tension through three forms of possible racial boundary-blurring: skin color differentiation within a nominal racial group, multiracial self-definition stretching across groups or substituting for any, and the new social identities and medical treatments made possible through the genomics revolution. These are not the only forms of racial boundary-blurring in the United

States, but they press the meaning of racial identity in ways that are potentially most transformative.[2] They also raise, in different ways, the normative question of how to balance individual rights of exit and collective betterment.

The rest of the chapter has four sections. In the first three, I provide evidence in turn about the three forms of boundary-blurring and explore the political or ideological contestation around each. In the final section, I suggest policies and practices that might sharpen or soften the contest between individual and collective goals.

SKIN COLOR DIFFERENTIATION

People with comparatively light skin have been advantaged for centuries, both within the United States and around the world. Before the Civil War, "mulattoes"[3] were more likely to be free than darker-skinned blacks.[4] Among free African Americans, mulattoes were more likely to own their own farms (Bodenhorn 2003) and had much greater wealth (Bodenhorn and Ruebeck 2007) than darker blacks. Lighter-skinned black Union soldiers in the Civil War held more skilled jobs and higher military ranks than their darker counterparts; they were taller (a measure of nutrition) and less likely to die in the war (Hochschild and Weaver 2003).

Social, economic, and classificatory differences perhaps deepened after emancipation. "Complexion homogamy"—the tendency for people to marry others of a similar color—prevailed, and by the turn of the twentieth century, light-skinned couples were wealthier (Bodenhorn 2006; see also Gatewood 1990; Hill 2000; Reuter 1917), more likely to be literate, and better situated to send their children to school (U. S. Bureau of the Census 1918).[5] Some state laws differentiated mulattoes from Negroes, and a few granted the former legal privileges denied to blacks. Initially at the behest of Congress and later for ostensibly scientific purposes, the United States Census Bureau enumerated mulattoes separately from Negroes (though always as a subset of them) for all but one of the eight censuses from 1850 through 1920 (Hochschild and Powell 2008). A steady stream of cases required courts to classify people on the margin as white or nonwhite, lawfully married or illegally miscegenated, and permitted to inherit property or debarred from it. (Cases include *Ex parte Shahid* 1913; *United States v. Bhagat Singh Thind* 1923; *Rhinelander v. Rhinelander* 1927. Analyses include Elliott 1999; Golub 2005; Gross 1998; Haney López 1996; Harris 1993; Lewis and Ardizzone 2001; Pascoe 1996; Zackodnik 2001.)

African Americans have, of course, long been aware of social and economic differences associated with skin color. The author Charles Chesnutt (1898) described a (apocryphal?) group whose

> purpose was to establish and maintain correct social standards among a people whose social condition presented almost unlimited room for improvement. By accident, combined perhaps with some natural affinity, the society consisted of

> individuals who were, generally speaking, more white than black. Some envious outsider made the suggestion that no one was eligible for membership who was not white enough to show blue veins. (55)

To the charge that they discriminated against darker-skinned people, the society's members "declared that character and culture were the only things considered; and that if most of their members were light-colored, it was because such persons, as a rule, had had better opportunities to qualify themselves for membership." Others deplored the pernicious effect of snobbery and self-denial that seemed to lie behind some blacks' desire to look and act more like whites. For example, sociologist Kelly Miller (1919) wrote in response to a book seeking to distinguish light-skinned mulattoes from darker blacks, "The dual caste system is undemocratic and un-Christian enough; to add a third would be inexcusable compounding of iniquity" (220).

After roughly 1930, discussion of skin color was withdrawn from the public into the private arena. State laws and the census bureau no longer distinguished mulattoes from blacks, and the term faded from the popular media. The black-oriented *Pittsburgh Courier,* for example, referred to "mulattoes" an average of thirty-nine times a year from its inception in 1923 through 1941, eighteen times a year from 1942 through 1963, and only four times a year from 1964 to the present. The white-oriented press followed the same pattern (Hochschild, Powell, and Weaver 2008).

But disparities between dark- and light-skinned African Americans persist. Many scholars, using the seven national surveys (from 1961 to 1994) with a skin-color measure as well as local or more opportunistic surveys, have found the same pattern (Hochschild and Weaver 2007). For example, in the 1979–1980 National Survey of Black Americans (NSBA; Jackson and Gurin 1987), a gap of almost two years separated the schooling of the darkest and lightest among the respondents (p = .01), and dark-skinned blacks earned less than 70 percent as much as light-skinned blacks, during a year in which black families' mean income was 63 percent of that of white families (p = .001). (See also Allen, Telles, and Hunter 2000; Bowman, Muhammad, and Ifatunji 2004; Cotton 1997; Hill 2000; Keith and Herring 1991; Krieger, Sidney, and Coakley 1998; Seltzer and Smith 1991.) Other features of life are affected by skin color, as well. To choose only one example, among 90,500 recently incarcerated felons in Georgia, the prison sentences of blacks with the lightest skins averaged 2,590 days (about as many as whites), medium-skinned blacks received a longer sentence (about the average for blacks), and the dark-skinned got hit with almost a full year more of incarceration than the light-skinned (Burch 2005). Dark-skinned blacks are more likely to receive a death sentence for capital crimes than their light-skinned counterparts (Eberhardt et al. 2006).

Statistical controls reduce the direct relationships between skin color and these outcomes, as one would expect given the long history of skin color effects. That is, characteristics such as parents' education and income or growing up in the suburbs or in the north, may themselves be indicators of inherited skin tone (dis)advantage.

However, controls do not eliminate skin color disparities; there remains a direct effect of appearance, such that blacks with darker complexions are treated worse than blacks with lighter complexions, net of other factors.

Skin tone differences are also related to holding elective political office, which means that dark-skinned blacks not only suffer from more discrimination but also enjoy less descriptive representation with which to contest it. From 1865 to the present, among all 116 African Americans elected to statewide or national office, 40 percent have been light-skinned—more than twice the proportion of the black population. Conversely, only 16 percent of elite black elected officials have been dark-skinned, compared with almost 40 percent of African Americans (coding by author; population data from NSBA). When faced with a choice between two fictitious African American candidates in a survey experiment, white respondents, including racial liberals, were more likely to "vote" for the lighter-skinned than the darker-skinned candidate. They were also more likely to describe the former as more intelligent, experienced, and trustworthy (though not more hardworking) (Weaver forthcoming 2011).

Finally, skin color is still related to both blacks' and whites' assessments of black individuals. Light skin is connected with more positive evaluations of photographs (Maddox and Gray 2002), perceived desirability as a dating or marital partner (Hunter 2002), children's appeal to would-be adoptive parents (Kennedy 2003), and black employees' standing with a black boss (Russell, Wilson, and Hall 1992). Except for a brief period in the 1960s, magazines and movies have always preferred light-skinned models and actresses (Leslie 1995; Woodson 1994. For more evidence on similar patterns, see Hochschild 2005 and Hochschild and Weaver 2007).

These findings are completely uncontroversial among analysts, who have consistently found that, with occasional exceptions focusing on concerns about racial authenticity, dark-skinned blacks suffer from discrimination in addition to that of race. Even beyond the hierarchy of nominal racial groups—white over black—skin color remains deeply implicated in racial inequality and poverty, perhaps as much as it was around the turn of the twentieth century.

Nevertheless, there is no political movement around disparities associated with skin color. The disadvantages attendant on dark skin are widely known, commonly acknowledged—and politically inert. In the array of groups mobilized around demands for rights, justice, and equality, no group advocates against colorism. Few cases have been filed in court against discrimination by "color" (from the 1964 Civil Rights Act), and none has succeeded. Survey data reveal the same lack of mobilization. The seven national surveys with a skin color measure include 212 politically relevant items, asking about perceptions of discrimination and white hostility, group consciousness, and group nationalism. Dark-skinned blacks do indeed see more racism or have more group identity on thirty-five items (at $p = .05$ or better)—but light-skinned blacks see more racism on nine items and no skin color group does on the remaining 168. Even the statistically significant differences were substantively small. These are not impressive results.

My co-author and I call this discrepancy between strong impact in the social, economic, cultural, psychological, and electoral arenas on the one hand and virtually no impact on political opinions or activities on the other hand the "skin color paradox" (Hochschild and Weaver 2007). It is an empirical puzzle and perhaps a political surprise—but is it a moral problem? Should we be concerned that African Americans suffering from secondary as well as primary discrimination (Cohen 1999) have no public representation, no organized recourse, and no systematic way to fight it? Or should we be relieved, on the grounds that focusing on skin color discrimination against some would divert attention and resources away from the more urgent matter of racial discrimination against all blacks? This is the dilemma with which I started—activity that would distinguish blacks with dark skin from other African Americans and that might benefit them could arguably interfere with activism for the group as a whole. What would be good for individuals could harm the community. I return to this dilemma below.

MULTIRACIALISM

As the history and my own practice above of blurring mulattoes and light-skinned African Americans suggests, multiracialism is closely related to skin color differentiation. The two phenomena are not the same; dark- as well as light-skinned people can have racially mixed ancestry, and siblings can vary in their skin tones. More importantly, self-chosen multiracialism draws our attention to the possibility of cultural mixing or multiple group loyalties, whereas skin color differentiation has more to do with how a person is perceived than how that person identifies. As with light skin, Americans have both celebrated and deplored racial mixture. Focusing on people now perceived as whites, the French traveler J. Hector St. John de Crèvecoeur (1782) famously asked,

> What then is the American, this new man? . . . I could point out to you a family, whose grandfather was an Englishman, whose wife was Dutch, whose son married a French woman, and whose present four sons have now four wives of different nations. . . . Here individuals of all nations are melted into a new race of men, whose labours and posterity will one day cause great changes in the world. . . . The American is a new man, who acts upon new principles. (43–44)

Over a century later, Israel Zangwill's "Melting Pot" had expanded beyond the races of Europe: "There she lies, the great Melting Pot. . . . Ah, what a stirring and a seething! Celt and Latin, Slav and Teuton, Greek and Syrian,—black and yellow" (Zangwill 1909). But these were unusual European Americans; more commonly, race scientists deplored racial mixture and insisted that since blacks, whites, and Indians were different species, their offspring would be degenerate and sterile.

African Americans also ran the gamut of views, from Marcus Garvey's description of W. E. B. Du Bois as "a little Dutch, a little French, a little Negro . . . a monstrosity" (Garvey 1969 [1925], 310) to the optimistic claim that "the race problem in America would be on its way to a quick solution if the white and Negro families of the deep South who are linked by ties of blood would simply recognize that relation today" (Martin 1956).

As this latter comment envisions, multiracialism has reversed the trajectory of skin color differentiation; mixed-race individuals have moved from being almost invisible in the public arena to forming organized groups and making claims for public recognition and acceptance. From the 1930s through the 1990s, Americans officially maintained the fiction that each person has only one race. If people marked more than one race on census forms, the Census Bureau reallocated them into a single race. But by 2000, the federal Office of Management and Budget (OMB) as well as some states, universities, and other organizations began to permit people to identify as "multiracial" or to choose more than one race on registration forms and official documents. About 2.4 percent of the American population claimed more than one race on the 2000 census (not including people who identified as Hispanic and also chose a racial category).

That is a small percentage, but it has increased slightly in the subsequent decade (Hochschild, Weaver, and Burch forthcoming 2012). Analysts are uncertain about the trajectory of multiracial identity, pointing both to evidence that "changes in marital assimilation have taken on momentum of their own, that is, America's growing biracial population has fueled the growth of interracial marriages with whites," and to the fact that monoracial racial identity remains strong even among those who could choose to identify as mixed race (Qian and Lichter 2007, 68).

Multiracial identification may well increase over the next few decades. Marriage and child-bearing across races is legal everywhere in the United States, and it is rising in most population groups. Just over 20 percent of Asian husbands under age 30 and just under 20 percent of comparable black husbands have wives of a different race.[6] As of the early 2000s, among children under age eighteen living in married-couple families, 6.4 percent have parents of different races; the proportions have surely risen since then, as the 2010 census will show (Farley 2007; see also Fryer Jr. 2007). Perhaps most importantly, a relatively small proportion of interracial or interethnic marriages and children imply a substantial fraction of Americans with at least one member of their extended family not of the same race or ethnicity. By 2000, that was the situation for 22 percent of whites, half of blacks, 84 percent of Asians, and essentially all Native Americans.[7]

Beyond demographic change, people with a multiracial identity are putting pressure on organizations to recognize it. Adoption agencies, public schools, college student associations, hospitals treating infants with jaundice or adults needing bone marrow transplants are, correspondingly, newly attentive to issues of racial mixture; federal population data now routinely report such categories as "white only," "white in combination with other races," and so on. Federal agencies ranging from the Department of Education through the military now mandate some version of the option of "mark one or more" in their data collection. As the concept of multiracialism comes to be embedded in the fabric of American life, more and more people may move away from the one-drop conception of race into the view that racial boundaries blur and overlap.

So like the skin color paradox, the trajectory of multiracialism presents an intriguing set of empirical possibilities. But is it a moral issue? In my view, yes. It presents a painful irony, illustrated by the two epigraphs to this chapter. The defeat of anti-miscegenation laws created the space needed for multiracial claims on American society, but those claims conflict with the racial cohesion that enabled successful challenges to racist laws, such as those prohibiting interracial marriage. That is, opponents of racial mixing (social as well as marital) first promulgated one-drop laws and other constructs of racial purity in order to maintain segregation. Eventually most blacks also embraced the one-drop understanding of race and used it to develop a strong sense of group identity and deep perception of shared fates—sentiments that they then deployed to fight segregation and white-imposed racial purity. The Supreme Court's repudiation of anti-miscegenation laws in 1967 was a victory for the civil rights movement, but that decision (along with increased immigration) opened the door to mixed-race marriages and multiracial children. Some of those children and their parents now contest the one-drop understanding of race and are thereby perhaps undermining racial solidarity among those generally understood to be African Americans.

That, at any rate, has been the fear of the NAACP and other advocacy groups—that an individual's desire to connect with both parents or all ancestries and to choose an identity outside the five conventional nominal groups (black, white, Asian, Indian, and Latino) will undermine nonwhite groups' ability to fight persistent racial hierarchy in the United States. Such a desire is especially problematic for small groups, such as blacks, Asians, and Native Americans, since even if an equal number of whites and nonwhites chose a multiracial rather than a monoracial identity, the proportions lost to small groups would be much greater than the proportion lost to the very large group of European Americans.

Individual's choice of identity versus the shared benefits of group unity: we are back to the dilemma with which this chapter started. I will explore it further after considering the third form of racial boundary-blurring.

GENOMICS

The cellular revolution is just beginning:

> There is in biology at the moment a sense of barely contained expectations reminiscent of the physical sciences at the beginning of the 20th century. It is a feeling of

advancing into the unknown, and that where this advance will lead is both exciting and mysterious.... The analogy between 20th-century physics and 21st-century biology will continue, for both good and ill. (*Economist* 2007, 13)

The cliché of both good and ill accords well with the mix of dire warnings and excited revelations evident in the encounter between race and genomics.[8] That mix, in turn, is interwoven with the tension between focusing on distinctive individual traits or on group cohesion.

On the one hand, the study of genomics emphasizes the arbitrariness of nominal racial lines and fosters racial boundary-blurring, just as multiracialism does and as skin color differentiation could do. Early systematic studies of worldwide DNA samples showed that "classification into races has proved to be a futile exercise.... We can identify 'clusters' of populations ... [but] at no level can clusters be identified with races, since every level of clustering would determine a different partition and there is no biological reason to prefer a particular one" (Cavalli-Sforza, Menozzi, and Piazza 1994, 19). On the other hand, some recent studies have reinforced claims about group-based characteristics: "A team of scientists at the University of Utah has proposed that the unusual pattern of genetic diseases seen among Jews of central or northern European origin ... is the result of natural selection for enhanced intellectual ability" (Wade 2005). This is a potentially explosive line of research; as cognitive psychologist Steven Pinker observed in response, "It would be hard to overstate how politically incorrect this paper is" (ibid).

The branch of genomics that focuses on an individual's genetic genealogy can, correspondingly, emphasize either racial boundary-blurring and individual variation or the ties between individual and group. Henry Louis Gates provides an example of boundary-blurring. After testing his ancestry, he received a surprise: "As for my mitochondrial DNA, my mother's mother's mother's lineage? Would it be Yoruba, as I fervently hoped? ... A number of exact matches turned up, leading straight back to that African Kingdom called Northern Europe, to the genes of (among others) a female Ashkenazi Jew." His response: "I have the blues. Can I still have the blues?" (Gates Jr. 2006) Another tester was more succinct: "I expected Kunte Kinte, and got Lord De La Warr" (Kilgannon 2007).

Mika Stump exemplifies the opposite impact of DNA ancestry testing. She grew up in foster homes, knowing nothing of her roots except that she is black. "But a DNA test she took recently showed strong similarities between Stump's genetic code and the Mende and Temne people of Sierra Leone.... Now, 'I have a place where I can go back and say, "This is who I am; this is my home." That's something I never, ever expected to say'" (Willing 2006). Recreational genomics may be especially likely to confirm and deepen African Americans' solidarity with a single population group since it can fill in some of the ancestral links lost through enslavement. That, in any case, is the line taken by the increasing number of firms that market DNA testing to black Americans. I know of no systematic evidence showing how many people move toward boundary-loosening or tightening after learning about their genomic heritage; anecdotal evidence suggests both effects.

The medical arena is probably the most important locus for both research and normative evaluation—and it too provides evidence that people can use genomics either to reinforce individual distinctiveness or to strengthen racial boundaries. With regard to the latter, new DNA research is identifying biological bases of diseases, some of which are disproportionately found in particular nominal racial groups. For example, "Decode Genetics reported in 2006 that African Americans have twice the incidence of a gene on chromosome 8 that causes susceptibility to prostate cancer, and African American women tend to suffer from very aggressive forms of breast cancer" (Nathan 2007, 12). Or, "reports were published … of a genetic variant associated with heart disease that, in the one-quarter of Caucasians tested who carry two copies, increases the chance of a heart attack by more than 50%" (*Science* 2007, 821). Careful scientists caution that all diseases are powerfully shaped by environmental and personal factors as well as genetic tendencies, that these results remain tentative, and that it may be decades before prescriptive medical treatment emerges from these scientific breakthroughs. Still, a leading researcher admits to "a great deal of excitement, because everyone realizes the field is changing so fast" (ibid.).

Some African American researchers share the excitement on the grounds that if black patients are especially susceptible to a particular disease, they are also especially likely to gain from discoveries of treatment growing out of the new research. Howard University established the National Human Genome Center (NHGC) in 2001 in order to "explore the science of and teach the knowledge about DNA sequence variation and its interaction with the environment in the causality, prevention, and treatment of diseases common in African American and other African Diaspora populations." Its scientists investigate genetic associations with obesity, diabetes, hypertension, and metabolic syndrome, while remaining attentive to connections between genes and environment.

Other researchers, however, fear possible implications of the claim that individuals' shared membership in a given race makes them especially susceptible to certain diseases. Sociologist Troy Duster warns that "recent research in medicine and genetics makes it … crucial to resist actively the temptation to deploy racial categories as if immutable in nature and society." Genomics research "holds promise" for finding drugs for "special subpopulations that have some functional genetic markers." But doctors must not use phenotype or nominal racial categories as a shortcut to identifying those subpopulations: "Research [must] be conducted to find the markers that have the actual functional association with drug responsiveness—thus assuring that the drug be approved for everyone with those markers, regardless of their ancestry" (Duster 2005, 1050–51). Using race as a shortcut for determining possible susceptibility to a disease, in short, risks both over- and under-inclusion as well as loss of the recognition that racial categories are socially constructed. Duster and many others, including the staff of Howard University's NHGC, also worry that in the search for biological "causes" of disease, the array of social and environmental conditions that increase the likelihood of illness will be even more ignored than they are now.

Duster's deepest concern, however, is the possibility of research programs aiming to link "SNP patterns [single nucleotide polymorphisms, the tiny genetic unit underlying the new research breakthroughs] and searches for a biological basis for criminal behavior." That is, from using race as a shorthand for identifying populations at higher risk of a given disease, it is only a small step to using race as a shorthand for identifying populations with propensities to violence or sexual aggressiveness (ibid, 1051). And the race so identified will, he anticipates, be black. Another writer is more blunt: "We are ill-prepared to respond to the complex challenges posed by racial arguments bobbing in the unstoppable tide of genetic research" (Kohn 2006).

We face here another version of the potential conflict between what is good for an individual and what is good for the group to which individuals belong. Individuals will benefit from scientific research that yields treatments for their particular ailments; testimonials from African American patients and doctors expressing gratitude for the drug BiDil, which may be marketed only to blacks with a specific form of congestive heart failure, make that point poignantly clear. But nonwhite groups in which some of those people are members may be harmed by research that has the effect of reinforcing stereotypes that a given race is disproportionately violent, unintelligent, or diseased. Very few Americans have the scientific, political, or moral training to deal with the classic liberal/communitarian conundrum in such a novel and complex arena.[9]

We have, then, three arenas in which racial boundaries are being challenged in some fashion to the potential benefit of individual challengers but potential cost to their group. Skin color differentiation reveals degrees of disadvantage in being a black American based on appearance. Awareness of that point might lead to challenges against secondary as well as, or at the expense of, primary marginalization. Multiracialism demonstrates degrees of monoracial identity, including thinking of oneself as not only a "mixed-race black" but as somehow distinct from the one-drop racial construction. Genomic science could go in many directions—reifying racial boundaries by re-biologizing race, helping to find cures for diseases to which specific groups are especially prone, loosening a person's identification with a particular racial category, deepening a person's commitment to African (or Irish) heritage, and perhaps others. In each boundary challenge, one can find the moral conundrum of liberalism's individual autonomy potentially posed against communitarianism's pursuit of group benefits.

These are not merely philosophical puzzles. With regard to skin color, for example, as a child, Supreme Court Justice Clarence Thomas became "harden[ed] ... against some of the most successful products of his race. To him, most blacks of a lighter hue were snobs, the self-anointed superior class of the race who considered themselves a cut above dark-skinned blacks with broad noses and thick lips, like himself. This class-and-color consciousness [is] not uncommon in the South" (Merida and Fletcher 2007, 49). With regard to multiracialism, the chair of the US Civil Rights Commission testified against its official recognition on the grounds that he could "see a whole host of light-skinned black Americans running for the door the minute they have another choice" (Fletcher 1993). These are raw emotions with potentially powerful political impact.

Beneath each dispute is the question of how necessary racial solidarity is at this point in American history. If racial hierarchy persists, a strong group identity remains essential to fight it; if racial hierarchy is becoming attenuated, strong group identity is less essential and may itself stand in the way of individual fulfillment just as racism did. I cannot hope to resolve that larger question here, but it underlies the empirical material presented above as well as the more philosophical discussion to which I now turn.

ANALYZING THE RELATIONSHIP BETWEEN INDIVIDUAL AND GROUP

To a strong liberal, there is no moral tension between individual autonomy and the good of the group because there is no group separate from the aggregation of individuals who comprise it: "Liberal individualism has the implication that a group has no value over and above its value to its members (and to other people outside it)" (Barry 2001, 123). If people associated with a group decide, separately or together, correctly or not, that their lives would be enhanced by moving away from the group's ambit, they have every right to do so.

Liberal individualism, of course, has an old and honored lineage in American political thought. My favorite exemplar is the curmudgeonly loner, Henry David Thoreau, whose aphorisms cover the classic ground at great speed: "What a man thinks of himself, that is what determines, or rather indicates, his fate" (Thoreau 1971 [1854], 7). Or "I know of no more encouraging fact than the unquestionable ability of man to elevate his life by conscious endeavor" (ibid, 90). And most pertinently, "It is never too late to give up our prejudices. No way of thinking or doing, however ancient, can be trusted without proof.... Old deeds for old people, and new deeds for new" (ibid, 8). In this view, not keeping pace with one's companions because one hears a different drummer is not apostasy, but the sign of autonomy and a newly open context, whether racial or otherwise.

Communitarians might agree that moral tension between individual and group-based goods is not necessary but for the opposite reason: group membership is so central to an individual's self-understanding that one can barely imagine a psychologically healthy person who separates him or herself from a group identity. That identity may not be elected; we are not "bound only by the ends and roles we choose for ourselves," argues Michael Sandel. Rather, we can "sometimes be obligated to fulfill certain ends we have not chosen—ends given ... by our identities as members of families, peoples, cultures, or traditions" (Sandel 1994, 1768).

In 1915 Horace Kallen provided the political lesson arising from a similar assertion: "Men may change their clothes, their politics, their wives, their religions, their philosophies.... They cannot change their grandfathers. Jews or Poles or Anglo-Saxons, in order to cease being Jews or Poles or Anglo-Saxons, would have to cease to be, while they could cease to be citizens or church members or carpenters or lawyers without ceasing to be." As a consequence, "a possible great and truly democratic commonwealth ... would be ... a democracy of nationalities, cooperating voluntarily and autonomously through common institutions in the enterprise of self-realization through the perfection of men according to their kind" (Kallen 1998 [1924], 114–16). From this vantage point, if each individual chooses not to keep pace with his companions, it does not matter whether he hears a different drummer; the result will be alienation for the person and loss of standing for the group.

Liberalism, Communitarianism, and Skin Color

Consider these positions in the realm of skin color differentiation. Blacks "must weigh concern over the respectability and legitimization of black communities in the eyes of dominant groups against concern over the well-being of those most vulnerable in our communities, as they struggle against very public, stigmatizing issues" (Cohen 1999, 15). Since the 1930s, almost all black leaders and advocacy organizations have given primacy to the fight against dominant groups, and have considered it necessary to submerge stigmatizing issues such as the social status and material benefits associated with light skin. Representative Eleanor Holmes Norton (D-DC) articulates both this need for solidarity as well as the Sandelian sense that racial identity is not, in any case, a matter of choice: "I sit here as a light skin black woman, and I sit here to tell you that I am black. That people who are my color in this country will always be treated as black.... We who are black have got to say 'Look, we are people of color, and we are readily identified. Any discrimination against one of us is discrimination against another'" (Norton 1997, 260).

But at least some of "those most vulnerable" find this view unpersuasive since Norton's ideal is not always met and alternatives might do better. As a feature story about a state legislator put it, "Well-to-do, fair-skinned kids ... weren't allowed to play with him and they regularly taunted him about his color, Jones says.... 'That's been a dominant force in my life. Having lived through those experiences gave me the desire to fight for the disadvantaged'" (Robinson-English 2005, 154). From a liberal vantage point, the black community has no right to ask Mr. Jones to remain silent about this injustice—that is, to sacrifice himself and others like him to the presumed common good of the race.

Liberalism, Communitarianism, and Multiracialism

Applying the contending claims of liberalism and communitarianism to multiracialism is more complicated, since the issue has a mixed ideological valence. Consider three distinct positions.

Liberalism maps easily onto multiracialism: each American has the right to celebrate all—or none—of his or her heritages. A college student illustrates this view:

> I did not have a problem until someone said, "How can you consider yourself interracial? You are black!" [The professor in my Black Awareness class said,] "You can't be both." So I said, "Well, I am both, you can't tell me I am not." So he said, "If there was a war, blacks are on one side and whites are on the other side, which side would you go on?" I said, "Probably neither, because I would have to choose between my father and my mother and I don't have a favorite." (Brown 2001, 47)

In the political arena, the Libertarian party was one of the few advocacy groups to support multiracial organizations' efforts to change the 2000 census, sometimes to the consternation of those who saw multiracialism as a new civil rights claim. Then-Speaker of the House of Representatives Newt Gingrich agreed, on the grounds that recognition of racial boundary-blurring was a step toward eliminating racial identifications altogether:

> America is too big and too diverse to categorize each and every one of us into four rigid racial categories.... Ideally, I believe we should have one box on federal forms that simply reads: "American." But, if that is not possible at this point ... we should allow them [Americans] the option of selecting the category "multiracial," which I believe will be an important step toward transcending racial division and reflecting the melting pot which is America. (Gingrich 1997, 661–62)

Roger Clegg, of the Center for Equal Opportunity, was more blunt: "Insisting that people embrace a racial identity is bad for civil-rights progress and, therefore, bad for civil-rights enforcement. Discrimination is more likely to occur in a society in which people have strong racial identities and an us-them mentality" (Clegg 2002, 2). So liberals support the legitimacy of multiracial self-definition, either as a component of autonomy and self-fashioning or as a step toward abolishing the constraining attachment to racial groups.

Communitarians appear on both sides of this issue. On the one hand, multiracial advocates often see themselves as promoting the next wave of civil rights activism for a newly identified group (DaCosta 2007). The mission statement of the Association of Multiethnic Americans (AMEA) reads, in its entirety, "To educate and advocate on behalf of multiethnic individuals and families by collaborating with others to eradicate all forms of discrimination" (AMEA 1997).[10] In this view, just as group solidarity enabled blacks to fight back against discriminatory treatment, so shared multiracial identity can provide the strength to contest harms to people who blur racial boundaries. The problem, therefore, lies not in the claims of multiracial individuals versus monoracial groups but rather in the unwillingness of older groups to make room for newer ones.

On the other hand, traditional civil rights groups sometimes mistrust multiracials' motives (they are seeking a "mulatto escape hatch") and reject their claim that multiracialism represents a new civil rights frontier. They speak on behalf of older group identities. Consider the claim in the second epigraph of this chapter: "A multiracial classification might disaggregate the apparent numbers of members of discrete minority groups, diluting benefits to which they are entitled as a protective class under civil rights laws and under the Constitution itself. In our quest for self-identification, we must take care not to recreate, reinforce or even expand the caste system we are all trying so hard to overcome" (McDougall 1997, 308). The 1993 and 1997 Congressional hearings saw dozens of similar statements; only one member of the Congressional Black Caucus supported the multiracial advocates.

Thus liberalism is readily associated with support for multiracial identity, albeit for several, perhaps contradictory, reasons. Communitarianism, however, may provide support for either side of the debate, depending on how one defines the relevant community. Whether this mix of views will permit reconciliation of individual and group interests in this realm remains to be seen.

Liberalism, Communitarianism, and DNA

Ideological complexities multiply when we consider genomic science. For one thing, the field of research is very new and is changing at an astonishing pace, so any analysis today may be outdated within a few years. In addition, the genomics revolution has many facets; I have focused on medicine and social identity, but it may also revolutionize the criminal justice system, markets for health insurance, adoption practices, food production, energy consumption, and other arenas. Most importantly here, the politics and social norms around genomics are even newer and more fluid than the science; it is simply not clear what a classic liberal (or communitarian) should want. Nevertheless, we can at least begin to apply the template of liberalism and communitarianism to see where the moral issues in this form of racial boundary-blurring are likely to arise.

A classic liberal will celebrate medical advances that lead to individualized medicine. Improved health makes it easier for a person to be autonomous, pursue goals, and retain dignity. More abstractly, individualized medical diagnoses and treatment will undermine arbitrary nominal racial (or other) groupings. A liberal will endorse the form of recreational genomics that identifies each person's particular ancestry for similar reasons; dissolving arbitrary group identifications into unique configurations will reinforce the view that everyone is distinctive and fundamentally equal.

A classic liberal might, however, worry that, intentionally or not, the genomics revolution will enhance the role of biology in explaining human behavior. The new knowledge may foster the views that biology is destiny, that choices matter little (or are themselves genetically determined), or that public attention to social environments is not needed. Those views could undermine Americans' commitment to autonomy, self-definition, and choice. A liberal might also note that historical eras in which race was largely understood in biological terms have been periods in which individual rights took a back seat to eugenics, concerns for racial purity, and the search for objective scientific evidence of what a race really "is."

A communitarian could also be deeply divided about genomics. An advocate for the black community must celebrate genomic medical advances that save the lives of African Americans suffering from distinct forms of heart failure, hypertension, or sickle-cell anemia. She will also presumably feel deep sympathy for the person who finds a cherished heritage in a particular African tribe or region. If, as Sandel claims, we are born into families, peoples, cultures, and traditions that do and ought to shape our sense of who we are, then a technology that connects individuals to peoples and cultures will help to make them more fully human.

But a communitarian advocate for the black community must also be wary of potential directions for genomics research. If scientists today find the genetic component of prostate cancer, tomorrow they may claim to find the genetic component of violence. And if the former is more commonly observed in people who are conventionally understood to be black, then why not the latter? In short, communitarians are likely to applaud those features of genomics research that underscore black solidarity but to deplore those features that could reinforce stereotypes of blacks as a group. That is a perfectly sensible moral stance, but it may be difficult to sustain it in research agendas, public discourse, and political decision making.

Each form of racial boundary-blurring, then, presents the conundrum of individual versus group benefit, or more generally, liberal versus communitarian values. As one moves from skin color differentiation through multiracialism to genomics, however, the relationship between particular features of the issue and the broad philosophical principles gets more complicated. As a consequence, the politics will be more uncertain, coalitions more fraught, and the moral valence of different positions more open to interpretation.

SHARPENING, OR SOFTENING, THE CONFLICT BETWEEN INDIVIDUALS AND GROUPS

So far, I have been arguing as though there is a fixed level of conflict between what is good for the individual and for the group, or between liberalism and communitarianism. If that is the case, our task can only be to understand the conflict and perhaps choose one side. But what if the level of conflict can vary, can be softened or sharpened by new understandings or new policy choices?

Policies and Practices that Heighten Conflict

Most remedies for the primary marginalization of racial hierarchy do nothing to help, or even exacerbate, the secondary marginalization of colorism.[11] Increases in the number of black Americans in elite electoral office over the past few decades have largely reinforced the relatively advantaged position of light-skinned blacks. Affirmative action policies similarly benefit light-skinned African Americans disproportionately through the mechanism of class. Black students, some of whom have benefited from affirmative action policies, in a sample of twenty-eight highly selective colleges and universities are less affluent than their fellow students who are white or Asian—but they are much more affluent and have parents with much higher levels of education and occupational status than most blacks (Massey et al. 2003, 41–44). Of the 808 black students in this sample, interviewers rated 30 percent as light-skinned (0 to 4 on a 10-point scale), 39 percent as medium-toned (5 and 6), and 31 percent as dark-skinned (7 through 10) (analyses of National Longitudinal Survey of Freshmen by Vesla Weaver [Massey and Charles 2006]). That compares with a national population rated by interviewers in 1980 as 17 percent light-skinned, 45 percent medium, and 38 percent dark (analyses of National Survey of Black Americans by Vesla Weaver [Jackson and Gurin 1987]). In short, black students at highly selective institutions of higher education are disproportionately well-off and light-skinned. Attending a highly selective college or university, in turn, gives a boost to graduates' income and occupational status as well as giving them an excellent education (Bowen and Bok 1998). Thus affirmative action arguably has the unintended consequence of exacerbating secondary while reducing primary marginalization. Recognizing this fact, one person evaluating college applications proposed that applicants be required to include photographs so that there could be affirmative action to offset colorism. But the article was written in a tone of heavy irony without any serious intention behind it (Fleming 2003), and I know of no other proposal or effort to offset dark-skin disadvantage.

Policies that require combining all mixed-race people into a single "multiracial" category for reporting purposes to public agencies could also exacerbate the tension between liberal rights and communitarian good of the group. If people can be disaggregated into their single-race groups (through a "mark one or more" recording system), it is possible to count blacks (or any other group) in the traditional way for purposes of regulation, litigation over discrimination, or other social goals. Thus individuals can choose to register their identity as mixed, while groups can simultaneously count those individuals in their numbers. But if all mixed-race people are combined into a single "multiracial" category, there is no way to determine how many of them claim black (or any other group) ancestry so they are "lost" to any monoracial group. That probably doesn't matter so long as tiny fractions of the American population identify as multiracial, but if 10 percent, 20 percent, or more come to do so—a demographically plausible possibility—then litigation or policies such as majority-minority districting or affirmative action could be powerfully affected.

In short, some public policies and social norms—even or especially those that seek to promote racial equality—could make the issue of racial boundary-blurring into an even more difficult conundrum than it already is. But they need not do so, a point to which I now turn.

Policies and Practices that Soften Conflict

Some policies to combat racial discrimination, such as majority-minority districting or laws against employment discrimination, could diminish tension between individual and group if African Americans come to believe that racial hierarchy is itself being reduced. Once blacks no longer feel the need for group solidarity in order to offset threats from the dominant white order, then racial boundary-blurring will no longer be so fraught and individuals will feel more able to identify with the group that matters most to them or to none at all.

Direct attention to skin color discrimination might help to ease the difficulties of dark-skinned blacks and reduce the temptations of light-skinned blacks to take advantage of their ascribed status. If African Americans and their supporters came to see colorism as part of the broader racial struggle against the norm of "whiter is better," then political mobilization around skin color discrimination could become part of, rather than a distraction from, political mobilization around racism. In terms of the conundrum of this chapter, the old cliché that "sunlight is the best disinfectant" suggests that direct attention to skin color discrimination will not only benefit individuals who get tangled in this sensitive and fraught issue, it will also reinforce blacks' mutual shared regard.

The tensions around multiracialism could be softened in several ways. First, despite some advocates' contentions (DaCosta 2007), the best of the scanty evidence available suggests that multiracial identity is itself fluid and highly contextual (Hochschild, Weaver, and Burch forthcoming 2012, chapter 3; Farley 2004). Surveys show that people often identify as multiracial in one setting but monoracial in another (Harris and Sim 2002) or that people who report parents of two races do not necessarily identify as multiracial (analysis of *Washington Post* Poll #2001-WPH012: Race and Ethnicity [Kaiser Family Foundation, *Washington Post,* and Harvard University 2006] by Vesla Weaver). Most people in racially mixed marriages do not see themselves or their children as torn between two identities (*Washington Post,* Kaiser Family Foundation, and Harvard University School of Public Health 2001). If most multiracials come to feel reasonably comfortable in shifting identities as the context warrants, then the conflict between individual and group can be lessened if not eliminated.

Perhaps multiracialism is even a step toward the form of assimilation that encourages people not to abandon racial identity but rather to hold several identities more or less strongly as the situation warrants. President Barack Obama on different occasions has described himself as a "mutt," new American, or black. He has invoked his white mother and grandmother to reassure whites that he can understand and share many of their views while always identifying with African Americans. Whether he or other Americans can sustain this complex set of maneuvers remains to be seen, but his behaviors suggest a pathway toward eventually reducing the gap between the self and the group.

With regard to genomics, federal or state governments might regulate some lines of especially volatile research. The National Institutes of Health could refuse federal funding for the study of genetic components of violence if that research also sought to determine the distribution of such components by race. The Federal Drug Administration could deny licenses for race-specific medications. Federal agencies could require that a nontrivial proportion of each research grant include analysis of the social or environmental causes, as well as the genetic components, of the disease in question. And of course, legislators, administrators, scientists, and other leaders can continue to seek to create a culture in which genomics research is completely dissociated from the eugenics and racial science of a century ago. At least one researcher has given up work purportedly showing that genetic changes in brain size, which might be linked to intelligence, had occurred over the past few thousand years in most continents but not sub-Saharan Africa. "It's too controversial," he reported (Regalado 2006, A1). Whether one sees this withdrawal as caving to political correctness or as a belated recognition that some research programs are socially inappropriate, it seems clear that cultural norms can exert some control over the most polarizing aspects of genomics research.

Perhaps the venerable philosophical debate between individual rights and community good will gain new traction in the developing arena of racial boundary-blurring. But perhaps Americans can figure out how to simultaneously ameliorate racial discrimination *and* skin color inequities, respond to the desire of multiracials to honor all of their progenitors *and* honor group solidarity, fulfill the promises *and* control the dangers of genomics research. We can predict with some confidence that racial boundaries will shift over the next century and the meaning of race will change; how Americans handle those developments will help to determine whether the United States can finally crawl out from under the weight of its racial history.

NOTES

1. My deep thanks for Traci Burch and Vesla Weaver for their comments on this chapter, and for their co-authorship of the larger book project of which this is an offshoot.

2. In the larger project, my co-authors and I also consider the ways in which immigration challenges the boundaries around American racial and ethnic groups.

3. After a first usage in quotations, I continue to use terms that were appropriate to the era being discussed even if they grate on contemporary ears—thus mulattoes, negroes or Negroes, half-breeds, and so on. I do not endorse use of these terms in the present, but constant quotations are annoying to a reader.

4. With a roughly similar history and impact, skin color differentiation occurs in practically all racial or ethnic groups, not just African Americans. Given space constraints, however, this chapter considers only the latter. For the same reason, I focus on American blacks in the discussions of multiracialism and the social consequences of genomic science.

5. Some of these data are measured in terms of mulattoes rather than light-skinned blacks, but during the early twentieth century, the two terms are roughly synonymous.

6. The proportion of Asian intermarriages is declining, especially among women (probably as a result of immigration), and the proportion of black intermarriages is rising, especially among men. Two-thirds of American Indians marry outside their race, and those proportions are still rising (Farley 2007).

7. Joshua Goldstein, communication with author January 29, 2009; for background, see Goldstein 1999. If Hispanic ethnicity were added to the model, the intermarriage rate would be higher and the kinship connection with someone of a different group would be dramatically higher.

8. "The branch of genetics that studies organisms in terms of their genomes (their full DNA sequences)" according to the simplest definition (WordNet® 3.0 n.d.).

9. The lab science is moving very fast, and the society and polity have not begun to catch up. Nor have social scientific analyses; for initial forays, see Condit and Bates 2005; Fullwiley 2007, 2008; Koenig, Lee, and Richardson 2008; Panofsky 2006; Phillips, Odunlami, and Bonham 2007; Singer et al. 2010.

10. AMEA argued against "mark one or more" on the census because it permitted disaggregation into the traditional single-race groups. It discusses multiracials' "common characteristic" on its website, and uses the language of "a racial group" to describe multiracials (AMEA 2001).

11. Much of this and the next paragraph comes from Hochschild and Weaver 2007.

REFERENCES

Allen, Walter, Edward Telles, and Margaret Hunter. 2000. "Skin Color, Income and Education: A Comparison of African Americans and Mexican Americans." *National Journal of Sociology* 12(1): 129–80.

AMEA. 1997. "Welcome to AMEA." Accessed November 14, 2004. http://www.ameasite.org.

———. 2001. "AMEA Responds to Multiracial Census Data." Accessed July 23, 2007. http://www.ameasite.org/census/031201censusdata.asp.

Barry, Brian. 2001. *Culture and Equality: An Egalitarian Critique of Multiculturalism.* Cambridge, MA: Harvard University Press.

Bennett, Claudette. 2003. "Exploring the Consistency of Race Reporting in Census 2000 and the Census Quality Survey." Paper presented at the annual meeting of the American Statistical Association, San Francisco, CA, August 3–7.

Bodenhorn, Howard. 2003. "The Complexion Gap: The Economic Consequences of Color among Free African Americans in the Rural Antebellum South." In *Advances in Agricultural Economic History,* vol. 2, edited by Kyle Kauffman, 41–73. Amsterdam: Elsevier Science North-Holland.

———. 2006. "Colorism, Complexion Homogamy, and Household Wealth: Some Historical Evidence." *AEA Papers and Proceedings* 96(2): 256–60.

Bodenhorn, Howard, and Christopher Ruebeck. 2007. "Colorism and African-American Wealth: Evidence from the Nineteenth-Century South." *Journal of Population Economics* 20(3): 599–620.

Bowen, William, and Derek Bok. 1998. *The Shape of the River: Long-Term Consequences of Considering Race in College and University Admissions.* Princeton, NJ: Princeton University Press.

Bowman, Phillip, Ray Muhammad, and Mosi Ifatunji. 2004. "Skin Tone, Class, and Racial Attitudes Among African Americans." In *Skin Deep: How Race and Complexion Matter in the "Color-Blind" Era,* edited by Cedric Herring, Verna Keith, and Hayward Horton, 128–58. Chicago, IL: University of Illinois Press.

Brown, Ursula. 2001. *The Interracial Experience: Growing Up Black/White Racially Mixed in the United States.* Westport, CT: Praeger.

Burch, Traci. 2005. *Skin Color and the Criminal Justice System: Beyond Black-White Disparities in Sentencing.* Cambridge, MA: Harvard University, Department of Government.

Cavalli-Sforza, L. Luca, Paolo Menozzi, and Alberto Piazza. 1994. *The History and Geography of Human Genes.* Princeton, NJ: Princeton University Press.

Chesnutt, Charles. 1898. "The Wife of His Youth." *The Atlantic Monthly* 82: 55–61.

Clegg, Roger. 2002. "Statement Re: Racial and Ethnic Data Collection by Government Agencies." US Commission on Civil Rights.

Cohen, Cathy. 1999. *The Boundaries of Blackness: AIDS and the Breakdown of Black Politics.* Chicago, IL: University of Chicago Press.

Condit, Celeste, and Benjamin Bates. 2005. "How Lay People Respond to Messages about Genetics, Health, and Race." *Clinical Genetics* 68(2): 97–105.

Cotton, Jeremiah. 1997. "Color or Culture? Wage Differences among Non-Hispanic Black Males, Hispanic Black Males, and Hispanic White Males." In *African Americans and Post-Industrial Labor Markets,* edited by James Stewart, 61–75. New Brunswick, NJ: Transaction.

DaCosta, Kimberly. 2007. *Making Multiracials: State, Family, and Market in the Redrawing of the Color Line.* Stanford, CA: Stanford University Press.

de Crèvecoeur, J. Hector St. John. 1997[1782]. *Letters from an American Farmer.* Oxford and New York: Oxford University Press.

Duster, Troy. 2005. "Race and Reification in Science." *Science,* February 18, 1050–51.

Eberhardt, Jennifer, Paul Davies, Valerie Purdie-Vaughns, and Sheri Johnson. 2006. "Looking Deathworthy: Perceived Stereotypicality of Black Defendants Predicts Capital-Sentencing Outcomes." *Psychological Science* 17(5): 383–86.

Economist. 2007. "Biology's Big Bang." June 16, 13.

Elliott, Michael. 1999. "Telling the Difference: Nineteenth-Century Legal Narratives of Racial Taxonomy." *Law & Social Inquiry* 24(3): 611–36.

Ex parte Shahid. 205 F. 812 (E.D.S.C. 1913).

Farley, Reynolds. 2004. "Identifying with Multiple Races: A Social Movement that Succeeded but Failed?" In *The Changing Terrain of Race and Ethnicity,* edited by Maria Krysan and Amanda Lewis, 123–48. New York: Russell Sage Foundation.

———. 2007. "The Declining Multiple Race Population of the United States: The American Community Survey, 2000 to 2005." Paper presented at the annual meeting of the Population Association of America, New York City: March 29–31.

Fleming, Bruce. 2003. "Not Affirmative, Sir: A Well-Meaning Admissions Board's Absurd Reality." *Washington Post,* February 16.

Fletcher, Arthur. 1993. Testimony at Hearings of the Review of Federal Measurements of Race and Ethnicity. Washington, DC: Committee on Post Office and Civil Service, U.S. House of Representatives.

Fryer, Roland Jr. 2007. "Guess Who's Been Coming to Dinner? Trends in Interracial Marriage over the 20th Century." *Journal of Economic Perspectives* 21(2): 71–90.

Fullwiley, Duana. 2007. "The Molecularization of Race: Institutionalizing Human Difference in Pharmacogenetics Practice." *Science as Culture* 16(1): 1–30.

———. 2008. "The Biologistical Construction of Race: 'Admixture' Technology and The New Genetic Medicine." *Social Studies of Science* 38(5): 695–735.

Garvey, Marcus. 1969 [1925]. *Philosophy and Opinions of Marcus Garvey, Vol. I and II.* New York: Atheneum.

Gates, Henry Louis Jr. 2006. "My Yiddishe Mama." *Wall Street Journal,* February 1, A14.

Gatewood, Willard. 1990. *Aristocrats of Color: The Black Elite, 1880–1920.* Bloomington, IN: Indiana University Press.

Gingrich, Newt. 1997. Testimony at Hearings on Federal Measures of Race and Ethnicity and the Implications for the 2000 Census. Washington, DC: Committee on Post Office and Civil Service, U.S. House of Representatives.

Goldstein, Joshua. 1999. "Kinship Networks That Cross Racial Lines: The Exception or the Rule?" *Demography* 36(3): 399–407.

Golub, Mark. 2005. "*Plessy* as 'Passing': Judicial Responses to Ambiguously Raced Bodies in *Plessy v. Ferguson.*" *Law & Society Review* 39(3): 563–600.

Gross, Ariela. 1998. "Litigating Whiteness: Trials of Racial Determination in the Nineteenth-Century South." *Yale Law Journal* 108(1): 109–88.

Haney López, Ian. 1996. *White by Law: The Legal Construction of Race.* New York: NYU Press.

Harris, Cheryl. 1993. "Whiteness as Property." *Harvard Law Review* 106: 1709–95.

Harris, David, and Jeremiah Sim. 2002. "Who Is Multiracial? Assessing the Complexity of Lived Race." *American Sociological Review* 67(4): 614–27.

Hill, Mark. 2000. "Color Differences in the Socioeconomic Status of African American Men: Results of a Longitudinal Study." *Social Forces* 78(4): 1437–60.

Hirschman, Albert. 1970. *Exit, Voice, and Loyalty.* Cambridge, MA: Harvard University Press.

Hochschild, Jennifer. 2005. "From Nominal to Ordinal: Reconceiving Racial and Ethnic Hierarchy in the United States." In *The Politics of Democratic Inclusion,* edited by Christina Wolbrecht and Rodney Hero, 19–44. Philadelphia, PA: Temple University Press.

Hochschild, Jennifer, and Brenna Marea Powell. 2008. "Racial Reorganization and the United States Census 1850–1930: Mulattoes, Half-Breeds, Mixed Parentage, Hindoos, and the Mexican Race." *Studies in American Political Development* 22(1): 59–96.

Hochschild, Jennifer, Brenna Marea Powell, and Vesla Weaver. 2008. "Political Discourse on Racial Mixture: American Newspapers, 1865 to 1970." Paper presented at the annual meeting of the Policy History Conference, St. Louis, Missouri, May 20–June 1.

Hochschild, Jennifer, and Vesla Weaver. 2003. "From Race to Color: Does Skin Color Hierarchy Transform Racial Classification?" Paper presented at Harvard Color Lines Conference, Segregation and Integration in America's Present and Future, Cambridge, MA, August 30–September 1.

———. 2007. "The Skin Color Paradox and the American Racial Order." *Social Forces* 86(2): 643–70.

Hochschild, Jennifer, Vesla Weaver, and Traci Burch. Forthcoming 2012. *Transforming the American Racial Order: Immigration, Multiracialism, DNA, and Cohort Change.* Princeton, NJ: Princeton University Press.

Hunter, Margaret. 2002. "'If You're Light You're Alright': Light Skin Color as Social Capital for Women of Color." *Gender & Society* 16(2): 175–93.

Jackson, James, and Gerald Gurin. 1987. National Survey of Black Americans, 1979–1980 (machine-readable data file). 1st ICPSR ed. Ann Arbor, MI: University of Michigan, Inter-University Consortium for Political and Social Research.

Kaiser Family Foundation, *Washington Post,* and Harvard University. 2006. ICR/Harvard/Kaiser/*Washington Post* Poll #2001-WPH012: Race and Ethnicity.

Kallen, Horace. 1998 [1924]. *Culture and Democracy in the United States.* New Brunswick, NJ: Transaction Publishers.

Keenan, Kevin. 1996. "Skin Tones and Physical Features of Blacks in Magazine Advertisements." *Journalism and Mass Communication Quarterly* 73(4): 905–12.

Keith, Verna, and Cedric Herring. 1991. "Skin Tone and Stratification in the Black Community." *American Journal of Sociology* 97(3): 760–78.

Kennedy, Randall. 2003. *Interracial Intimacies: Sex, Marriage, Identity, and Adoption.* New York: Pantheon.

Kilgannon, Corey. 2007. "At a Harlem Reunion, a Rancher from Missouri Meets his 'DNA Cousin.'" *New York Times.* March 14: E3.

Koenig, Barbara, Sandra Soo-jin Lee, and Sarah Richardson, eds. 2008. *Revisiting Race in a Genomic Age.* New Brunswick, NJ: Rutgers University Press.

Kohn, Marek. 2006. "The Racist Undercurrent in the Tide of Genetic Research." *Guardian,* January 17, 26.

Krieger, Nancy, Stephen Sidney, and Eugenie Coakley. 1998. "Racial Discrimination and Skin Color in the CARDIA Study: Implications for Public Health Research." *American Journal of Public Health* 88(9): 1308–13.

Leslie, Michael. 1995. "Slow Fade to ?: Advertising in *Ebony Magazine,* 1957–1989." *Journalism and Mass Communication Quarterly* 72(2): 416–35.

Lewis, Earl, and Heidi Ardizzone. 2001. *Love on Trial: An American Scandal in Black and White.* New York: Norton.

Maddox, Keith, and Stephanie Gray. 2002. "Cognitive Representations of Black Americans: Re-exploring the Role of Skin Tone." *Personality and Social Psychology Bulletin* 28(2): 250–59.

Martin, Louis. 1956. "Dope and Data." *Chicago Defender.* April 21, 9.

Massey, Douglas and Camille Charles. 2006. "National Longitudinal Survey of Freshmen." Princeton University, Office of Population Research, http://nlsf.princeton.edu/index.htm.

Massey, Douglas, Camille Charles, Garvey Lundy, and Mary Fischer. 2003. *The Source of the River: The Social Origins of Freshmen at America's Selective Colleges and Universities.* Princeton, NJ: Princeton University Press.

McBride, James. 1996. *The Color of Water: A Black Man's Tribute to His White Mother.* New York: Riverhead Books, Putnam.

McDougall, Harold. 1997. Statement at Hearings on Federal Measures of Race and Ethnicity and the Implications for the 2000 Census. Washington, DC: Committee on Post Office and Civil Service, U.S. House of Representatives.

Merida, Kevin, and Michael Fletcher. 2007. *Supreme Discomfort: The Divided Soul of Clarence Thomas.* New York: Doubleday.

Miller, Kelly. 1919. "Review of the Mulatto in the United States." *American Journal of Sociology* 25(2): 218–24.

Nathan, David. 2007. *The Cancer Treatment Revolution.* Hoboken, NJ: John Wiley & Sons.

Norton, Eleanor Holmes. 1997. Statement at Hearings on Federal Measures of Race and Ethnicity and the Implications for the 2000 Census. Washington, DC: Committee on Post Office and Civil Service, U.S. House of Representatives, 259–61.

Panofsky, Aaron. 2006. "Why Behavior Genetics Can't Kick Its Race Problem." Paper presented at the annual meeting of the American Sociological Association, Montreal, Quebec, Canada, August 10.

Pascoe, Peggy. 1996. "Miscegenation Law, Court Cases, and Ideologies of 'Race' in Twentieth-Century America." *Journal of American History* 83(1): 44–69.

Phillips, Elizabeth, Adebola Odunlami, and Vence Bonham. 2007. "Mixed Race: Understanding Difference in the Genome Era." *Social Forces* 86(2): 795–820.

Qian, Zhenchao, and Daniel Lichter. 2007. "Social Boundaries and Marital Assimilation: Interpreting Trends in Racial and Ethnic Intermarriage." *American Sociological Review* 72(1): 68–94.

Regalado, Antonio. 2006. "Scientist's Study of Brain Genes Sparks a Backlash." *Wall Street Journal,* June 16, A1.

Reuter, Edward. 1917. "The Superiority of the Mulatto." *American Journal of Sociology* 23(1): 83–106.

Rhinelander v. Rhinelander. 1927. 219 N.Y.S. 548 (N.Y. App. Div. 1927).

Robinson-English, Tracey. 2005. "Statehouse Clout." *Ebony,* April: 150–54.

Russell, Kathy, Midge Wilson, and Ronald Hall. 1992. *The Color Complex: The Politics of Skin Color among African Americans.* New York: Harcourt Brace Jovanovich.

Sandel, Michael. 1994. "Book Review: Political Liberalism." *Harvard Law Review* 107: 1765–96.

Science. 2007. "Closing the Net on Common Disease Genes." May 11, 820–22.

Seltzer, Richard, and Robert Smith. 1991. "Color Differences in the Afro-American Community and the Differences They Make." *Journal of Black Studies* 21(3): 279–86.

Singer, Eleanor. 2010. "The Effect of Question Framing and Response Options on the Relationship between Racial Attitudes and Beliefs about Genes as Causes of Behavior." *Public Opinion Quarterly* 74(3): 460–76.

Thoreau, Henry David. 1971 (1854). *Walden.* Princeton, NJ: Princeton University Press.

US Bureau of the Census. 1918. "Negro Population 1790–1915." Washington, DC: Government Printing Office.

United States v. Bhagat Singh Thind. 204. (261 US 1923).

Wade, Nicholas. 2005. "Researchers Say Intelligence and Diseases May Be Linked in Ashkenazic Genes." *New York Times,* June 3, A21.

Washington Post, Kaiser Family Foundation, and Harvard University School of Public Health. 2001. Survey of Biracial Couples.

Weaver, Vesla. Forthcoming 2011. "The Electoral Consequences of Skin Color: The 'Hidden' Side of Race in Politics." *Political Behavior.*

Willing, Richard. 2006. "DNA Rewrites History for African-Americans." *USA Today,* February 1.

Woodson, Laureece. 1994. "An Examination of Colorism and Variables Related to the Phenomenon of African-American and Mainstream Magazine Advertising." College Park, MD: University of Maryland, College of Journalism. M.A. Thesis.

WordNet® 3.0. n.d. "Genomics."

Zackodnik, Teresa. 2001. "Fixing the Color Line: The Mulatto, Southern Courts, and Racial Identity." *American Quarterly* 53(3): 420–51.

Zangwill, Israel. 1909. *The Melting-Pot.* New York: Macmillan.

Chapter 5

On Ethics and Economics

Kenneth J. Arrow, Kristen Renwick Monroe, and Nicholas Lampros

Conversation

Kristen Monroe (KM): Could we you tell us a little bit about yourself and how you ended up doing the kind of work you do—more precisely, how you came to economics, how you came to develop social choice theory, and how you got interested in ethics at all?

Kenneth Arrow (KA): To say that I'm interested in ethics is probably just to say that I'm a human being. You're brought up full of ethical concerns; these things are just part and parcel of life. I think the real thing is whether you think about ethical concerns rather than just accepting them. Will you raise the level of consciousness? Of course, if you raise the level of consciousness, there's some kind of dilemma. You find yourself with something that doesn't seem quite as straightforward as you thought. This is a very general observation. It doesn't speak to right and wrong, which are part and parcel of my past attitudes. This sometimes leads to counter-intuitive things like trying to take advantage of immediate situations. I was, for whatever reason, concerned from a pretty early age about issues like poverty, the inequality of income, and topics like racial barriers, which of course were very common in the 1930s; they were quite overt. Now we have a whole set of techniques in economics for detecting whether racial discrimination exists or not. You didn't need any of that in the 1930s; everybody was perfectly open about it. Everybody just said there were certain jobs, certain places to live depending on skin color. There was the racial question, the fact that there were large inequalities of income—a point that came up in election politics. This took the form for many students, and to some extent myself, of being radical on social activities of one kind or another.

On top of that, there were also what I call the practical failures, the efficiency failures of the economy, which were manifested in a 25 percent unemployment rate, for example. These were things that, apart from justice, made you feel that the system wasn't functioning properly. So that's the background for me, a background that of course was not in any way unique to me. Many students were brought up in this sort of milieu. But a lot of this evaporated in the post–World War II prosperity.

I also had the habit of trying to get things into logical order. That's my talent. I like to connect things that seem to belong to different realms. I see the parallels. I'm rather good at that. As Louis Pasteur once said, "Chance favors the prepared mind." I was rather trained—I trained myself and was trained by others—to say, "There's the logical consequence." One of these came up in the post-war period. I was involved, among other things, in working for the RAND Corporation, which was worried about the possibility of nuclear annihilation and our conflict with the Soviet Union, which I felt was personally very important. In spite of my leftist views, in many ways I took very early the opinion that the Soviet Union was a tyranny and was not the beacon of justice that many of my fellow students felt it was. Of course, many things came to light both during and after the war that made the situation even worse than I had thought it was.

Of course, one of the questions that came up was, "Well, what do we mean by the United States' interests?" The United States is an abstraction. The country is composed of a lot of people with different interests, just as the Soviet Union is. I attempted to address this question, which was raised to me by Olaf Helmer at the RAND Corporation, and meet it with the economics that I had been trained in. Because there had been an economic literature on, "when is a policy justified?" Let's say we're discussing imposing tariffs or some kind of taxation policy. When is it justified? A lot of very able people had written on this question. And I said, "Oh well, economists have discussed it." So my friend at RAND said, "Well, why don't you explain it?" As soon as I tried to explain it, I realized that there were some fundamental problems with the explanations that were in the literature. Rather unfortunately from my point of view, what came out was an impossibility result. I wasn't looking for impossibility; I was looking for an answer. I tried a number of answers and none of them worked. When I thought about what I meant by saying they "didn't work," I realized I meant there were certain implicit criteria in my mind, or I think in general in people's minds, for finding a good way of putting people's different preferences together to form a social preference. Then I realized it couldn't be done under the rules I was working with, rules that I thought were very reasonable and that I thought were very appealing to most people.

That experience lead to my social choice theory, or my impossibility theorem about the impossibility of social choice. In a way, it was kind of a setback to any ethical consideration. Of course, I did have the general economist's conviction that values ultimately inhere in people. They don't come from the outside. I realized as I read more philosophical literature that this proposition is by no means accepted by philosophers, particularly liberal philosophers, particularly ones that I'm most sympathetic with, who are writing down what they think the good life is. Rawls, for example—I'm skipping in time, of course, because Rawls published a good deal later—argued that majority voting doesn't have any particular significance. Things are either good or they're not good; the fact that the majority of people agree or not is not terribly relevant.

I thought it was a bit shocking because I had this inherent view that the aim of a society is not to have values imposed but to represent the (actual) values of the society. So this gave rise to the other questions: How do you represent the values? How were the values formed? These were questions that I found very difficult. I thought they were very important, but I didn't have anything to say about them. Having already shown impossibility, I wasn't in the mood to create new impossibilities, so I preferred not to say anything and I busied myself with what I thought economists could say about various things, such as what we could say about risk-bearing. I regarded the task of economic policy as to remove the obstacles to the free expression of people's preferences. If the market isn't doing it, then there must be other ways of doing it. This gives rise to the whole discussion about market failures and government expenditures. Well, this problem keeps on popping up. It's one thing that's led me back to somewhat more fundamental discussions of the more general question of provisions for the future. When you provide for the future, especially the long-run future, markets clearly don't work. They don't work because a lot of the people whose interests we're concerned with aren't even here. They can't express themselves in the markets. This is a question that's bothered me for a long time, almost from the beginning of my career. But I've always put it aside. Now I'm coming back to it. This is a concern dramatized by the climate-change issue. But in many ways this question of markets and the interests of future generations exists in other forms; the climate-change issue is not the only example.

There is a literature called optimal savings that is in a broader sense about provision for the future. You find almost everybody's written on it. It gets at some point into ethical questions. There was a classic paper in 1928 by Frank Ramsey, the English philosopher who dabbled in economics a little bit. His was a very strong statement about our responsibilities for the future. According to Frank Ramsey, even if you analyze very deeply, it is just taken for granted that future people count the same as present people. He had an ingenious method that side-stepped some of the issues. At least it appeared to do so. But as you look more closely, you realize he didn't really avoid the issues. You just didn't notice that at first. In fact, in other essays on different subjects, Ramsey expressed the conventional point of view that the future doesn't look like the present. It's like looking through a telescope; it's further away, so it's not as important. There's a dilemma of this kind that people are still struggling with. There's a new and important work on climate change that came out last year by the British, under the leadership of Nicholas Lord Stern. This work essentially took this old English attitude that the future people count like the present people. I'm exaggerating a little bit, but that's more or less it. Do you really mean that? I look in other ways at the logical implications of that position and say, "Do you really want to build your models on that assumption? Do these satisfy you?"

Another problem that has struck me—and which more recently has been the focus of my research—has been the problem of direct connections among people. That is, how does a consideration of one person's welfare affect another's? One way I have approached this topic in the past is through the question of income distribution. I referred a few minutes ago to the individualistic basis. I don't like to have values imposed by philosophers. So then how do we explain, for example, what justifies some sort of intervention in income distribution? Well, it's got to be based on individuals. Rich individuals feel they have to give some of their income away. Of course, there is some evidence that this actually happens. I won't go into it in great detail here, but there is, for example, a literature on voting for schools. People without children vote for schools; they vote to tax themselves for schools. Rich people who send their children to private schools vote for public school expenditures. Things like that. So there is some evidence on this question, and in a sense you can reduce the issue to individual preferences. I'm less and less satisfied with this; I'm more and more convinced that this isn't the whole story, that people are connected with each other in some intrinsic ways. Coming back to the time preference: if the people who are here now are really part of our community, but people who are far away in time—or in space for that matter—aren't as much part of our community, then somehow the sense of felt responsibility is less.

In the chapter I did for this book, I was stressing the family. People haven't discussed the family so much because in a sense it hasn't been regarded as problematic. It's not that people neglect the family. The family is so dominant people just take it for granted. But of course the family is now becoming more and more problematic. Illegitimacy rates are much, much higher than they were even thirty years ago. Society is undergoing a very rapid transformation compared to what the situation was years ago. The question of responsibility of parents has changed. But I'm more concerned with the responsibility that's been socialized in a way. Social security, Medicare, things like that. There are a lot of problems with that, but having the state get involved in issues like this does mean that certain parts of the problem are being addressed in a certain way. But the relationship of parents to children, the model of the indissoluble marriage with children, this model doesn't really fit much anymore. So the question of relations with children, responsibilities for children, has become much more problematic. Of course this mixes in with ethical questions. It's also mixed in with just ordinary descriptive questions, with what does happen. We have some evidence—considerable evidence I think—that children are adversely affected by shifts in family patterns but have no good answers as to what our responsibility is, given these shifts.

The state or the church or some other institution has always been a bit of a guardian. Perhaps its role is more than a bit. That certainly has been claimed. Certainly, it's been recognized implicitly that this role of institutions—such as the state—is a matter of public concern. It's recognized that other people have the right to interfere, that other people are acting through the state, that they have the right to interfere in these personal family relationships. (So the police are called in to investigate instances of child abuse, for example.) Presumably most of the time there's no conflict between private desires and public desires, so this problem isn't acute. But I think it is becoming a significant problem in this world. There are also the responsibilities of children through other mechanisms like education, and we have some problems in that respect also. The question is whether you can replace personal relationships with impersonal institutions. That's a major question here. So there's a whole complex set of issues for economists here that concern our ties to others. Even the description, even just predicting what the economy will do, depends in part on the ties we have to relatives and on our response to these social ties. This is both an economic and an ethical question. I hope that answers your question about how I got interested in ethical issues.

BIBLIOGRAPHY

Ramsey, Frank P. 1928. "A Mathematical Theory of Saving." *The Economic Journal* 38(152): 543–59.

Rawls, John. 1971. *A Theory of Justice.* Cambridge, MA: Harvard University Press.

Stern, Nicholas. 2006. "The Stern Review Report on the Economics of Climate Change." London: HM Treasury.

The Responsibility of Parents to Children

Kenneth J. Arrow

Investigation

The ethical obligations of parents to children belong to an intellectual realm that does not seem to me to have been very thoroughly explored. In secular discourse, especially that associated with economic and, more generally, consequentialist thinking, all individuals, including children, have rights and obligations induced by the need to preserve the benefits of society, but there is no special emphasis on the parent-child relation. Even religious discourse, though emphatic on the obligations of children to parents, places little emphasis on the reciprocal obligations. In fact, of course, a great deal of social activity, including economic activity, consists of parental support and maintenance of children. Moreover, the future adult life of children is profoundly affected by the family environment, so parental behavior becomes a serious externality for children, to use the economists' jargon. In fact, the legal systems of most countries recognize these problems, if only in an ad hoc manner.

I want to examine the questions of obligation that are raised by a consideration of these issues. I don't claim to have any definite answers. I do, however, want to illustrate that the issue is not purely imaginary, that there are real problems in the parent-child relation and in the legal system surrounding it.

Let me explain how I came to be interested in this question. It arose from considering some questions that are very much in the mainstream of economics. I realize that to many this will be a strange way of approaching the question announced in the title. Nevertheless, I believe the economic "take" on this social issue can be very useful. In any case, the reverse direction of influence is of great importance to economic understanding. Resources flow from parents to children both individually and collectively. The child is (typically) maintained by the parents up to some point and, to an extent, governed by both individual preferences and social norms enforced to some extent by the law. Inheritance and inter vivos gifts are significant factors in the distribution of income. These factors in turn are motives for saving and consumption. Further, society provides education largely, though not entirely, through collective means. The increasing concern over the environment, particularly with regard to its more permanent features such as climate change, are in part a concern for our children (and their children and so on).

My main scholarly concerns with this issue have only been through the problems of economic growth and especially the implications of environmental and resource degradation. But in the course of generally keeping up with the literature, a number of issues have bothered me.

The first was a minor issue in the controversies over the book by Robert Fogel and Stanley Engerman (1974), which dealt with the economics of slavery. Among many other courageous calculations, they attempted to measure the degree to which slaves were exploited. They had, on the one hand, an estimate of the productivity of a slave and, on the other, estimates of the consumptions of slave children and adults. Slaves started being productive at some given age (sixteen, as I recall). They discounted the productivities during working years to birth and the consumptions also to birth and compared them to find the degree of exploitation (the extent to which rewards fell below the competitive level, which would cause the two to be equal). One critic argued that this was fallacious; we should compare, he held, the productivities and consumptions beyond sixteen and discount them to the beginning of working life. He argued that this was the standard method in national income accounting, especially in discussing the distribution of personal income. This sounded to me like a real issue. In the case at hand, though, it was only mildly relevant.

The consumption of children before they begin work had to be accounted for somewhere. To be consistent with the critic's view, the consumption of children should be regarded as that of their parents. If, as was the case, the net reproduction rate was considerably greater than two, there would be some correction, but it would not make a great difference.

Years later, I undertook to give a course on the distribution of income, not because it was one of my fields of research but because I felt the subject was neglected in the curriculum. (It still is.) I found that distribution of income is presented by families, without any regard to the size of the family (to be sure, persons living alone are presented separately). There are two peculiarities of this presentation: (1) The family is treated as a unit, regardless of size. But our intuitions about income distribution are about the welfare of individuals. Clearly, for a family of a given income, there are more people affected in a larger family. (2) Since the income of each family has to be spread over the members, the per capita income is smaller.

This suggested to me that a useful way of representing the distribution of individual income would be to consider, for example, a family with an annual income of $50,000 that contains five members as five individuals each with an annual income of $10,000. There are obvious objections to this proposal, but they run in the direction of moving even farther from present practice. For one thing, small children have less need for space and food, so each child should count for less than one (so-called adult-equivalent scales). This raises some interesting questions in interpersonal comparability of income, which have not, I believe, been well addressed in the literature. For another, there are economies of scale in household consumption. It may not be true that "two can live as cheaply as one," but two can live together at the same utility levels for less than the sum of the costs of their living apart.

Apart from the detailed analyses of these cases (interesting and important in themselves), these two examples show the ambivalent relation between family and its members. The concept of the individual, which is so dominant in economic or legal discourse, gets a bit blurry when children are concerned. The statisticians have sensed an ambiguity in our understanding of this boundary between the individual and the social.

My aim is to propose the issue of a useful analytic representation of family structure, especially the role of children. The purpose is normative, rather than descriptive; I want to start a study of the formulation of a welfare economics for a world in which children play a role but not the same as that of adults. All I can do here is sketch some of the issues such a formulation will have to address.

SOCIAL ETHICS, RECIPROCITY, AND CHILDREN

Let us start with a quotation from the great eighteenth-century jurist, William Blackstone, in his famous *Commentaries on the Laws of England.* This sets forth the issues in a very clear way.

> The next, and most universal relation in nature, is ... that between parent and child.... The duty of parents to provide for the *maintenance* of their children, is a principle of natural law; an obligation, says Puffendorf, laid on them not only by nature herself, but by their own proper act, in bringing them into the world; for they would be in the highest manner injurious to their issue, if they only gave their children life, that they might afterwards see them perish.... The municipal laws in all well-regulated states have taken care to enforce this duty though Providence has done it more effectually than any laws, by implanting in the breast of every parent that ... insuperable degree of affection, which not even the deformity of person or mind, not even the wickedness, ingratitude, and rebellion of children can totally suppress or extinguish.... The power of parents over their children is derived from the former consideration, their duty: this authority being given them, partly to enable the parent more effectually to perform his duty, and partly as a recompence for his care and trouble in the faithful discharge of it. (Quoted from Mnookin and Weisberg 2000, 255–59)

Please note some of the relevant points: (1) there is a "natural" obligation in the parent-child relation, which is automatically recognized by the parent; (2) taking the action of giving birth to a child creates an obligation, since otherwise the child wouldn't exist (this is different from point 1); (3) society has the right (and the obligation?) to enforce the parental obligation, though usually the enforcement will be unnecessary, and (4) the parental obligation implies also power to the parents, for efficiency in discharging the obligation but also (very curiously) as a reward to the parents.

There is enough in these four points to sustain considerable analysis, more than I will engage here. Instead, with Blackstone's remarks in mind, I will consider the usual models for social analysis.

If one looks at the standard models that try to provide a normative foundation for social policy, we notice two common elements in many of them. These remarks apply equally to utilitarianism, to John Rawls's concept of justice as fairness (1971), and to contractarian views from Thomas Hobbes onward. One element is that of parity or reciprocity. The other is that of a mutual gain from social interaction.

Consider first reciprocity and parity. In the utilitarian ethical position, the criterion of social well-being is the sum of all individuals' utilities; all individuals enter symmetrically. The "original position" argument, as advanced by William Vickrey (1945), John Harsanyi (1955), and John Rawls (1971), gives some justification to this symmetry. It imagines individuals contracting for a social and institutional structure before they know their individual characteristics. This line of argument can be used to justify utilitarianism (Vickrey, Harsanyi) or other criteria (as in Rawls). In effect, redistribution or other ethical steps are derived as "insurance" under the uncertainty of knowing one's abilities and needs. This view implies that the contracting parties are in some sense drawn from the same universe of abilities and needs, even though they are not, strictly speaking, equal.

The second characteristic is that there are gains to all from the formation of a society. Thomas Hobbes, the fountainhead of the contractarian justification for social arrangements, vividly explained the advantages of society over the "state of nature" that would prevail in the absence of social order. Economists speak more prosaically of the gains from trade, which, in turn, may stem from specialization due to differing skills and needs or from increasing returns to scale.

If we take the conventional view of children as not-yet productive members of society, then they can provide neither insurance nor their parts of social gains. (This consideration is even more relevant to our obligations to generations still unborn.) The argument about insurance does have an important qualification. Children can be thought of as making an insurance contract over time; that is, they can, when they are grown and productive, take care of their now-retired parents. This function has certainly been real historically. It is now discharged collectively by social security systems, to the extent that they are not fully funded. However, I think it is clear that the net flow of resources is from parents to children if one includes everything (rearing, education, and inheritance).

These facts about children imply that most theories of justice have difficulty in accommodating the place of children. Rawls's ingenious attempt to study the ethics of regard for children and subsequent generations is very unsatisfactory; its inconsistencies have been noted by Arrow (1973) and Dasgupta (1974). Utilitarianism generally solves the problem by admitting that children's welfare enters the social maxim but with a discount factor.

Empirical works on the economics of the family, notably that of Becker (1981), have also faced this issue. Here, in effect, children are treated as durable consumer goods, though their welfare also enters into the family's welfare. This point of view would come closest to justifying treating the distribution of family income, uncorrected for family size, as the appropriate measure of income inequality.

SOCIAL ACTIONS AND INDIVIDUAL VALUES

Consider the problem of social choices in a general way. We have a society of individuals who must take a collective action. Economists may think of the outcome as the allocation of resources, but one could take a more general perspective. The point of view is that inherent in the usual economic analysis, that is, "methodological individualism." In the general social choice model, this means that the preferences of all individuals are taken into account in arriving at a social outcome.

Each individual is considered to have a set of values, that is, a set of preferences over social outcomes. Both the market and the political system (and their joint operation) can be taken as sets of rules that, for given technologies, determine the social outcomes (e.g., the distribution of consumption goods, jobs, ownership rights, the levels of public goods, and other objects of legislation and government policy) in a way that depends on individual value systems. The rules include spheres of individual decision and collective decision, which add up to a social outcome.

Hence, it is necessary for individuals to be able to express preferences. The bias of economic analysis is to assume that the individual represents his or her values through actions. These include voting, purchases and sales, or participation in the intermediate institutions of civil society, such as philanthropy and volunteer activities. It is also true that, in any system with an economic component, the possession of assets and productive abilities will cause an increase in the weight of one's individual value system on the social outcome.

How do children fit into this picture? Their ability to express preferences in actions (or other ways) is limited in two ways: (1) an inability to express preferences or to understand the consequences of actions, and (2) a lack of assets. (Both of these, of course, change with age.) To represent the true interests of a child, it appears that it is necessary to have an adult "trustee." The trustee's assets may also be diverted from the trustee's own interests to those of the child, as expressed in the quotation from Blackstone.

Evolution has suggested that the natural trustees are the parents. In effect, the Beckerian value of children as durable consumer goods can motivate a trusteeship relation. Obviously, this system has worked fairly well. Indeed, it is noteworthy that while the Bible is full of injunctions for children to honor and respect parents, there are none (to my knowledge) imposing any obligations on parents to respect and nurture children. The pessimistic interpretation is that children are the property of parents, who can do as they like with their children. But, more likely, the omission implies that parental love is so strong that good treatment of children need not be enjoined. The feeling of children for parents, per contra, needs reinforcement.

In short, there is a widely felt obligation that parents act as trustees for children. The concept of "obligation" is not widely spread in ethical and economic discourse about social arrangements. Rather, one talks about the achievement of happiness (even in the Declaration of Independence) or, as we frequently say, "utility" or even, as with Sen (1985), "functionings and capabilities." Another language talks of "rights." I now have come to believe that there is a category of "obligations" that cannot easily be reduced to either utilities or rights.

Though the recognition of the obligations of parents to children may be widespread in practice, it is far from universal, and therein lies the need for policy discussions. The need for at least some state intervention has received recognition in the law of the United States. Thus, in the case of *Prince v. Massachusetts,* the opinion of the Supreme Court held that,

> The state's authority over children's activities is broader than over like actions of adults.... A democratic society rests, for its continuance, upon the healthy, well-rounded growth of young people into full maturity as citizens, with all that implies.... Among evils most appropriate for such action are the crippling effects of child employment, more especially in public places, and the possible harms arising from other activities subject to all the diverse influences of the street. *It is too late now to doubt that legislation appropriately designed to reach such evils is within the state's police power, whether against the parent's claim to control of the child or one that religious scruples dictate contrary action.* (Emphasis added. Quoted from Mnookin and Weisberg 2000, 83)

Even the libertarians Milton and Rose Friedman (1962) have defended compulsory education laws on the grounds that the interests of parents differ from those of children.

SOME EVIDENCE

I here present some US data that bear on the extent to which there is a need to enforce the trusteeship obligations of parents. In the following, "children" are defined as those under eighteen. First of all, let us consider the income status of families with children. Certainly, it would be unfavorable for children to be brought up predominantly in households with little ability to provide for their maintenance and education. Here, the news is fairly good. Having children is positively associated with family income; 52 percent of children live in families with incomes in the top 40 percent of the income distribution; 54 percent of children live in families with incomes that are 200 percent of the poverty level or more.

However, the picture with regard to parental status is not good. In 2002, only 69 percent of children lived with two parents. The educational and future handicaps of being raised with one (or in 4 percent of the cases, no) parent have been heavily documented, even holding income constant. ("Parents" here include step-parents and adoptive parents.) Of course, single-parent households are also much poorer.

The figures on child neglect and abuse are also startling. Each year, about 1.3 percent of children are reported as being subject to abuse and neglect. This means that over eighteen years of childhood, 20 percent of children are, on the average, sufficiently abused or neglected to be reported. Other data, based on self-reports of adults, show that something like 12 or 13 percent of children have been severely abused. Severe child abuse is a major risk factor in adult depression.

The large number of single-parent households in the United States is driven to a great extent by births to unmarried couples. But an increasingly significant cause is the rise in the divorce rate, mostly due to the increased prevalence of laws permitting divorce on the instigation of a single party (unilateral divorce). It is also clear that unilateral divorce has significant impacts on the subsequent adult behavior of the children; they are less well educated and have lower family incomes. They also marry earlier but separate more often and have higher odds of adult suicide. None of these effects is very large, but in the population of the United States, there are considerable numbers of future adults at stake (Gruber 2004). It has also been shown that parental divorce has negative effects on the child's subsequent mental health. Some of this can be attributed to the emotional environment in families that subsequently choose divorce, but there is evidence that the divorce itself adds to the impact on mental health (Cherlin, Chase-Lansdale, and McRae 1998).

POLICY ISSUES

Clearly, the obligations to children require some attention. Analytically, we lack an adequate framework to discuss what obligations are legitimate and how they are to be traded off against other claims, say to utility. This lacuna is strongly exemplified in the discourse on population.

From the practical policy viewpoint, the evidence seems to be that our present perception of obligations is failing a significant fraction of the children. Increasing the role of the state may be called for in some cases, but we all know the problems there. The conflict shows up in every case of a child killed or gravely damaged by parents; there is a balancing of the state's rather crude and limited capabilities against the value of even an underperforming family.

I think, though I am far from confident, that our divorce laws have gone too far. Marriage per se is not a matter of great social interest; consenting adults should be free. But when children are involved, the situation changes considerably. There are, as the economist would say, externalities. However, I must admit that the most intractable part of single-parent households is the never-married. I have no good idea what can be done here.

REFERENCES

Arrow, Kenneth J. 1973. "Rawls's Principle of Just Savings." *Swedish Journal of Economics* 75: 323–35.

Becker, Gary. 1981. *Treatise on the Family.* Cambridge, MA: Harvard University Press.

Cherlin, Andrew J., P. Lindsay Chase-Lansdale, and Christine McRae. 1998. "Effects of Parental Divorce on Mental Health throughout the Life Course." *American Sociological Review* 63: 239–49.

Dasgupta, Partha S. 1974. "On Some Problems Arising from Professor Rawls's Conception of Distributive Justice." *Theory and Decision* 4: 325–44.

Fogel, Robert W., and Stanley L. Engerman. 1974. *Time on the Cross: The Economics of American Slavery.* Boston, MA: Little, Brown.

Friedman, Milton, and Rose Friedman. 1962. *Capitalism and Freedom.* Chicago: University of Chicago Press.

Gruber, Jonathan. 2004. "Is Making Divorce Easier Bad for Children? The Long-Run Implications of Unilateral Divorce." *Journal of Labor Economics* 22: 799–833.

Harsanyi, John C. 1955. "Cardinal Welfare, Individualistic Ethics, and Interpersonal Comparisons of Utility." *Journal of Political Economy* 63: 309–21.

Mnookin, Robert H., and D. Kelly Weisberg. 2000. *Child, Family, and State: Problems and Materials on Children and the Law.* 4th ed. Gaithersburg, NY: Aspen.

Rawls, John. 1971. *A Theory of Justice.* Cambridge, MA: Harvard University Press.

Sen, Amartya K. 1985. *Commodities and Capabilities.* Amsterdam: North Holland.

Vickrey, William. 1945. "Measuring Marginal Utility by Reactions to Risk." *Econometrica* 13: 319–33.

Chapter 6

Economics, Policy, and the Public Good

Thomas Schelling and Kristen Renwick Monroe

Conversation

Kristen Monroe (KM): Why don't you start by telling me just a little bit about yourself, your background, where you came from, and how you ended up becoming an economist?

Thomas Schelling (TS): I became an economist because I grew up during the Great Depression. The Great Depression didn't touch me because my father was a naval officer and we had an upper-middle-class income; Franklin Roosevelt, the president, had been the assistant secretary of the Navy and was very fond of the Navy, and didn't let anybody cut naval officers' salaries. But I decided that the worst problem in the world was economic depression, and economics probably had the solution. I later decided that war was worse than depression, but by then I had become an economist.

I got interested in nuclear issues because I spent five years in the Marshall Plan and related foreign aid programs, mainly engaging in negotiation with aid recipient countries or negotiation with the establishment of a European Payments Union. When I left the government in 1953 to go to Yale, I decided I would make bargaining theory my main interest, and I became aware of two things: one was that the thing I was doing was something like game theory, and the other was that probably the most important applications were in the relation between diplomacy and military power. Then I decided that maybe nuclear weapons policy was the most important application, and I thought that in connection with nuclear weapons everybody is an amateur—there is nobody who's been doing nuclear weapons policy for a career, and we've only had one use of nuclear weapons, which probably is not a model for what might happen. And some of my work attracted attention at the RAND Corporation. I spent a year there and learned enough about nuclear weapons, the logistics and the engineering and the technology of the nuclear weapons, and I also met the people who had done the most thinking about nuclear weapons policy. Then I began getting invited to the National War College to lecture, not on economics but on warfare, and eventually, I had a monopoly on the subject of warfare for the War College; I think seventeen years in a row I was the only one who lectured to them on war. And then I got especially interested in how to design weapons and delivery systems to make war less likely, or controllable if it happens, and I wrote a book on arms control, which became kind of a best seller as these things go.[1]

I think my training in economics was the best training I could have had, although since the 1950s, I have never published in an economics journal—except the *American Economic Review,* which wanted to publish my Nobel lecture. But I am published mainly in journals like *World Politics, Foreign Policy,* and things of that sort. A lot of people don't know that I am an economist.

KM: Did you have training in game theory?

TS: No, I didn't know any game theory until Duncan Luce and Howard Raiffa published a book in 1957 called *Games and Decisions.* I had already published two things that, in retrospect, I recognize as having been what you might call amateur non-technical game theory, and I think actually the Nobel Selection Committee considered them to be my most important publications even though there was no explicit game theory, but I really learned about game theory from the Luce and Raiffa book. I received an email asking whether I would be willing to go to Irvine to celebrate the fiftieth anniversary of that book. My response was that I'd go all the way to Singapore.

KM: I remember reading one of your books in graduate school. What do you think about the [Iraq] war that's going on?

TS: Oh geeze, I hate to think about it. I am very much impressed with my wife's reaction. She said, "You know, if Bush had a couple of housewives in his war cabinet, they would have made damn sure that he knew what was going to happen to the hospitals, the museums, the schools, and how to maintain law and order and keep everything from going to hell." All he had were gung-ho people, probably a lot of people who didn't know how to tell the President that he was fooling himself into thinking that this was a simple, short-run military conflict. And now people ask me what I would do, or what I would advise Barack Obama to say he would do, and I have to say I know so little about what's happening on the ground that I don't see what's the most respectable thing to do. And I think if I were Hillary Clinton or Barack Obama, I would say, "I'm not going to commit myself now. I will learn so much more when I am president; I want to leave open the decisions I will make when I am able to spend my full time concentrating on the war and talking to the best brains I can find."

KM: Have you been asked to advise for Obama?

TS: No, no one has asked me for advice.

KM: Your intellectual work has obviously had important policy implications. Have you ever wanted to move beyond that and become a policymaker yourself?

TS: Well, I spent a long time during the whole decade of the sixties being a consultant to the Pentagon and the State Department, even the White House. I spent five years in the foreign aid business. I found during the 1960s that I probably had more influence than if I had accepted a payroll position somewhere, because I was brought in to chair lots of committees on nuclear weapons policy. And I know that I was responsible for what was ultimately the hotline between Washington and Moscow, and I chaired a committee that I think was mainly responsible for ultimately the Anti-Ballistic Missile Treaty. So I had plenty of policy engagement, some of it successful. I quit that when President Nixon and Henry Kissinger invaded Cambodia in 1970. I led of group of Harvard faculty down to tell Henry Kissinger that we were no longer going to consult for him or his president; we no longer had any confidence in them.

KM: What was his reaction?

TS: He invited us to lunch; he didn't realize what we were going to do. I knew his military aid, who had been on a committee that I chaired, and I said a bunch of us want to come down and talk to Henry about the new events in the war. And then it was embarrassing because we were invited as Henry's lunch guests. I introduced us by saying that we were all his colleagues at Harvard and we were all totally dismayed with the invasion of Cambodia, and I said that I wanted to call on every one of them to tell him in their own words just how they feel about it. So I watched Henry Kissinger, who had been a close colleague of mine for nine years, sink deeper and deeper in his chair as all these people told him how disappointed they had become. Then he asked if he could go off the record and I said, "No, this is all on the record." I would say he looked gray in the face; he looked dreadful. And finally he said, "If I can't go off the record, let me just ask you to withhold your judgment for sixty days, and you'll see that we did the right thing." I went out and was asked by reporters, "How did he react?" and I said "I think we had a powerful influence on him." I later had to say that in their actions, I didn't see any sign that we'd had any influence on them, but I did write to him sixty days later and say, "I just polled; we unanimously agreed with what we told you sixty days ago." That was the end of my official consulting for any part of the government. I haven't gotten a dollar of pay from the federal government since then.

KM: Your work on race, have you ever done anything on that?

TS: Yeah, I published a lot of things on segregation and integration. In fact, one of the things that was mentioned by the Nobel Committee was something that I did in 1971, demonstrating how easily an integrated society can break up and become segregated, and how hard it is to take a segregated society and integrate it.[2] I actually conducted a war game back in 1961; I invented a kind of crisis game that I put on in Washington with people from the State, Defense, Treasury, White House, and CIA. Robert Kennedy, the attorney general, came up right afterwards and said, "You know, this could be very useful in coping with school integration issues and I would like you to come and talk to my brother, the president, about that." Well, that was Halloween 1963 and three weeks later the President was assassinated, and nothing ever came of it.

KM: Have you ever thought your work might be useful now with some of the issues we have with integration, not just racial integration but other types of integration?

TS: I think about it, but I am so preoccupied with global warming and nuclear weapons that I don't deal with integration. When I was still back at Harvard, we eventually wrote a book, a project on what would need to be decided if there were to be a Palestinian State—what would have to be decided *both* by the Palestinian State and Israel in terms of monetary and banking policy and taxation policy.[3] Because we'd still have Palestinians working in Israeli firms, and there would be issues of whether they'd have separate social security systems or the same retirement system, and what they're going to do about medical insurance. They would both depend a lot on excise taxes, and you don't want to have people driving across the border to buy gasoline in Palestine because it costs less than in Israel and vise versa. We got permission from the King of Jordan and Yasser Arafat and from an appropriate Israeli official to bring together Jordanians, Palestinians, and Israelis to work on the subject. We put together four different groups of Jordanians, Palestinians, and Israelis. They couldn't meet in the Middle East, but they were permitted to go to Cambridge, Massachusetts, and it was astonishing how well they got along; they just worked beautifully. And in the end we did publish a book of four chapters on what are going to be the essentials of Israeli-Palestinian relations when the Palestinians have a state. That was when we were optimistic that the Oslo

Accords were going to be adopted, but of course they never did. I was interested at the time that they could work so well together, but they couldn't work in Amman or either part of Jerusalem. There was a beautiful establishment up on the shore of Lake Galilee overlooking the Golan Heights, which was endowed by a woman who said, "I want this to be the place where Israel and Syria will negotiate a peace treaty." Well, they never got together for that peace treaty, but I used to think that might have been an appropriate place to hold meetings.

Lyndon Johnson had a terrific idea that was carried out and is a huge success. He got the idea for a scientific think tank to work on problems common to Soviet and American societies, or Eastern and Western societies, and they established the International Institute for Applied Systems Analysis, located near Vienna as a neutral territory. Howard Raiffa was the director of that institute, and it was sponsored by the Soviet Union, Ukraine, Hungary, Poland, Czechoslovakia, Canada, United States, Britain, France, and Germany. By now, it has lost Germany and Britain and picked up India and Pakistan and China. For all the decades before the Iron Curtain fell, it was about the only place in the world where Russians could go and meet with Americans on equal terms because it was a neutral think tank sponsored by five nations on each side. Then the Iron Curtain fell, but it was persuaded that now it might be able to do the same thing with India and Pakistan, or eventually Egypt and Israel, and since they hope to recruit Korea into it, maybe even North Korea. And I think that that could be a model for one way to get Israelis and Palestinians together with some Egyptians.

KM: How does the current Bush administration fit into that? They don't seem to be interested in brokering things.

TS: I don't think the Bush administration understands this. I think that if somebody could have persuaded Bush to do there what Lyndon Johnson did with the Cold War, it might have made a difference. Distinguished scientists from both sides could come together and work on common problems. I have given up on the Bush administration totally, but I think when we have a new president, even if it's John McCain, I am going to, through intermediaries, offer anything I can do to help, and maybe try to get people like Howard Raiffa to go talk with somebody about this institute in Vienna that serves such a fantastic purpose.

KM: What are the questions that you would still like to have answered if you had fifty more great years ahead of you and resources to set off on some research projects? What would be the things you would focus on? What would be some of the questions or topics you would like to address?

TS: I think that the problem of global warming is going to be with us for all those next fifty years.

KM: How would you focus research in that area?

TS: Well, a lot of it is going to be technological research; a lot of it is going to be what you might call geo-physical research. I think the most important uncertainty we have with respect to global warming is whether warming is going to release methane from under the oceans and permafrost to set off an indefinitely expanding warming process. I don't know how much research can find out about the likelihood that global warming may set in motion an unstoppable process. That is urgent research. Another one is that we know a probable economical way to store CO_2 out of fixed locations, like electric power plants, underground—maybe deep in the ocean but preferably underground. What we would need to know is what kind of geological formations may admit CO_2 and keep it indefinitely sustained. How can you test whether the CO_2 would be slowly released, and therefore it wouldn't do any good to put it down there, and where the locations are? China is building an electric power plant at the rate of at least one per week, all based on coal, and at least they should know where to locate them so that it's economical to capture the CO_2 and put it underground. It's going to require a huge amount of geological exploration and testing to find out what kinds of physical underground structures are suitable, and where we can find them and locate power plants so they don't have to pump CO_2 five hundred miles or one thousand miles to get there. There ought to be a huge international project of geological exploration as well as devising the appropriate technologies.

Otherwise, I think nobody has yet hit on a good idea for how to generate international cooperation. I don't think the Kyoto approach has done anything significant. I don't think what came out of Bali is going to be any good. I think that too many people, including good friends of mine, have a perfectionist idea that you can divide up all of the CO_2 to be emitted over the next fifty years among 190 countries with excellent monitoring and strong enough sanctions so that everyone will abide by the assigned quotas. I think there has never been any enforceable international scheme anything like the magnitude of what's going to be required here. People are going to have to think of alternative international approaches.

KM: Do you see any political volition to do this?

TS: No, but I have an idea and have been talking about it for several years. There is only one international system ever accomplished on a scale comparable to what is going to be required, and that's the North Atlantic Treaty Organization. Nothing was ever enforceable, the treaty was only two pages long, and it says nations commit to treat an attack on anyone as an attack on themselves. And it was a huge success. Fifteen nations recruited young men and put them into the army and trained them, and a lot of them sent them into Germany to be stationed. There was never any enforcement but at the outset, the United States was financing enough of the economies of the nations providing the troops that the United States had a lot of bargaining power. After three years, there was no more of that, yet it was a spectacularly successful system. It grew out of the Marshall Plan. The Marshall Plan put very high-ranking leaders of all the participating countries into diplomatic negotiations in Paris on how to divide up the aid because after the first year, the United States said, "You divide the aid, we won't divide it." There were people at the ministerial level on a first-name basis, haggling for months, cross-examining each other to reach a decision on how to divide up the next $5 billion.

KM: You were a part of that, weren't you? The Marshall Plan?

TS: Yeah, I was in Paris during all that. Not all of it, but a year and a half; then I was back in Washington, doing the same thing.

KM: It must have been an exciting time.

TS: It was, and I think that was the only time when everybody involved in the US government felt that we were doing something very good and very well.

NOTES

1. Schelling and Halperin 1961.
2. Schelling 1971.
3. Fischer et al. 1994.

REFERENCES

Fischer, Stanley, Leonard J. Hausman, Anna D. Karasik, and Thomas Schelling, eds. 1994. *Securing Peace in the Middle East: Project on Economic Transition.* Cambridge, MA: MIT Press.

Luce, R. Duncan, and Howard Raiffa. 1957. *Games and Decisions.* New York: John Wiley & Sons.

Schelling, Thomas C. 1971. "Dynamic Models of Segregation." *Journal of Mathematical Sociology* 1(2): 143–86.

Schelling, Thomas C., and Morton H. Halperin. 1961. *Strategy and Arms Control.* New York: The Twentieth Century Fund.

Climate Change

A Bundle of Uncertainties

Thomas Schelling

Investigation

My discussion of climate change is concerned with uncertainty. If this were just an American audience, I would have to explain that the concept of greenhouse gases may be, as my former president Bush used to say, "merely a theory," but then so is gravity merely a theory. Nobody understands gravity, but it's pretty well established that things are attracted to Earth if you let go of them.

If you shine some infrared light through a glass chamber full of carbon dioxide, less light comes out than went in because certain wavelengths in the infrared part of the spectrum are intercepted by carbon dioxide molecules. If you compare the difference between the light going in on one side and the light coming out on the other, it correlates with the rise in temperature of the gas that has intercepted the infrared radiation. That's been known for one hundred years—the basic idea that so-called greenhouse gases, carbon dioxide in particular, will absorb or intercept infrared radiation.

Now, the earth gets bathed in sunlight during the daytime, and it must get rid of that energy to avoid global warming. It gets rid of it in the form of infrared radiation. Therefore, what keeps the temperature of the atmosphere in balance is the earth's ability to get rid of the excess energy that comes in during the daytime. It's been understood for decades that the planet Mars, lacking any greenhouse gases, is too cold for water to exist on its surface as a liquid. It's been equally known that the planet Venus is so bathed in greenhouse gases that its surface is too hot for water to exist as a liquid. Earth has a nice concentration of greenhouse gases, which allows temperatures such that water can exist as a liquid. Without liquid water, there could be no life. All life on Earth depends on liquid water, and therefore while we may lament an excess of greenhouse gases in the atmosphere, we shouldn't forget that without carbon dioxide and associated greenhouse gases, we couldn't have any life on Earth.

Somebody persuaded the Supreme Court of the United States, about a year ago, to treat carbon dioxide as a pollutant. Now, it has to be remembered, carbon dioxide is what all green things "eat." Without carbon dioxide there would be no green grass, no trees, no vegetation. Carbohydrates all depend on carbon dioxide, which, through photosynthesis, energized by sunlight, causes green things to grow, so that herbivores can live and carnivores can live off the herbivores, all of which depend on carbon dioxide in the atmosphere. It's worth remembering that if we breathe carbon dioxide at many times the ordinary concentration in the atmosphere, there is no physical harm, not even to pregnant women or fetuses or small children. So it's important to keep in mind that what we call greenhouse gases are absolutely essential to life on this planet, and the only problem is, if there's too much, we may get climate changes that we don't like. (It is worth reflecting, just for perspective, that if the concentration of oxygen in the atmosphere doubled, that would be catastrophic. Any kind of fire would become a conflagration. Both gases are essential; too much of either can become dangerous.)

A word about the term "greenhouse": greenhouses don't work the way the greenhouse effect works. Greenhouses merely trap warm air: we build a glass enclosure, the sunlight warms the earth inside, and that warms the air inside; the function of the greenhouse glass is simply to keep the warm air from dissipating. That's not the way the greenhouse phenomenon works.

People refer to the greenhouse problem as a global warming problem; it's better to think of it as a climate change phenomenon. It isn't clear that every place will get warmer just because the average atmospheric surface temperature is going up. Some places may get cooler, some places may get sunnier, and some may get cloudier; some may get more precipitation and some may get less precipitation. Climate changes—and that has to be plural, not just *a* climate change but changes in climates everywhere—are still hard to predict.

To give an example: it's hard to predict what's going to happen high up in the mountains. Not many people live above 3,000 meters—Tibetans and Bolivians and a few such people—but huge numbers of people depend on what happens to the climate above 3,000 meters. That is because a huge part of the world's agriculture depends on moisture that falls in the form of snow in the wintertime and stays there as snow until late spring and early summer and then melts and comes down in rivers and is available for irrigation. Most agriculture in India, Pakistan, Bangladesh, Burma, China, Chile, Peru, Argentina, and California depends on precipitation in the form of snow that stays as snow until it's time to irrigate crops. If, instead of falling as snow, the precipitation falls as rain, it's pretty well wasted, or if it falls as snow and it melts too early, it's not available when the farmers need water. Therefore, what happens in the high mountains is important to agricultural productivity in most of the world.

Yet it's difficult for any of the current climate models to anticipate what's going to happen to climate above about 3,000 meters, though it's crucially important. So instead of a global warming problem, it's a problem of climate change, of which the driving force will be an increase in the average atmospheric surface temperature that will mainly manifest itself as greater warmth.

Now, among the uncertainties, the uncertainty that gets the most attention is what's going to happen to the average atmospheric surface temperature. That's just a beginning uncertainty! It is interesting to look at the estimate of how much temperature change there might be with a doubling of the concentration of greenhouse gases in the atmosphere—doubling vis-à-vis the so-called pre-industrial level of about 280 parts per million. The estimate, as of the late 1970s, of the US National Academy of Sciences was that a doubling of the concentration of CO_2 (at that time carbon dioxide was the only recognized so-called greenhouse gas) would lead to a change in temperature that could be anywhere between 1.5 degrees Celsius and 4.5 degrees Celsius. Now that's a huge range of uncertainty. It's a threefold difference: 1.5 degrees to 4.5 degrees Celsius. If I went to my doctor and he said, "Schelling, if you don't watch your diet, you're going to put on a lot of weight," and I said, "How much weight am I going to put on if I don't change my diet?" and my doctor said, "Oh, anywhere from 40 to 120 pounds," I'd say, "that's quite a range of uncertainty, Doctor!"

Now it's interesting that when the Intergovernmental Panel on Climate Change (IPCC) reviewed this estimate in 1992, it said, "We find no reason to change that estimate," and that is still about the going estimate. Sometimes it's referred to as maybe 2 degrees to 5 degrees. The question arises: why hasn't that estimate, the range of uncertainty, been reduced in the last thirty years? The amount of money spent on research into this question, the number of people dedicating their careers to studying climate change, is essentially now one thousand times what it was before the mid-1970s, and so the question arises: why hasn't that certainty been reduced?

In fact, as Marty Weitzman explained,[1] in some ways the uncertainty has increased. There are at least two reasons. One is that it takes a lot of self-confidence to offer an alternative estimate to the one that has been accepted for three decades, and nobody is quite ready to bet his career on a revised estimate. But more than that, there are some subjects—of which genetics and brain science are two examples—that upon examination turn out to be more complicated than people expected. Brain science now isn't anything like what it was thought to be twenty years ago, and genetics has changed drastically in the last couple of decades.

When I first got involved in this subject, the role of the oceans was just that of a huge cooling reservoir. We used the term "thermal inertia" to mean that because the specific heat of water is so much greater than that of the atmosphere, you can't get the air much warmer until you have warmed the surface of the ocean. The expectation was that it might take anywhere from a decade to half a century for the atmospheric temperature to get into what we refer to as equilibrium. Equilibrium is this idea: if, right now, we froze the concentration of greenhouse gases in the atmosphere—at somewhere around 385 parts per million—if we stabilized it there, it might take anywhere from ten to thirty, forty, or even fifty years for the air to stop getting warmer because until the surface of the ocean is in equilibrium with the air temperature, it will keep getting warmer until it reaches an equilibrium. Thirty years ago that was considered essentially the only role that oceans played. Now oceans are recognized as being active participants in the circulation of heat around the globe, and phenomena like the El Niño in the South Pacific Ocean are recognized to be the result of active oceanic participation in the circulation of heat around the globe.

Thirty years ago, clouds were ignored. Now it's recognized that clouds—depending on the sizes of the water droplets, on whether the clouds are ice crystals instead of water droplets, and on whether the clouds are over the oceans or over land masses—can either be absorbers of outgoing radiation or reflectors of incoming radiation. All of these things weren't known, or at least weren't appreciated, thirty or even twenty years ago.

Part of what I'm explaining is that this is a new subject. I participated, in the late 1970s, in two large energy studies. Some of you will remember that we had what was known as an energy crisis in the 1970s and early 1980s. I was in one project on nuclear energy: we had nuclear engineers, nuclear medical people, petroleum engineers, and economists—all sorts of people. We published a book, *Nuclear Power: Issues and Choices* (1977), a 400-page book. It had two pages on the greenhouse problem. It had a whole chapter on how bad coal might be for your lungs, but two pages out of 400 in a book about nuclear energy. If today you had a book on nuclear energy, one quarter of it would be on nuclear weapons proliferation, one quarter would be on how to dispose of nuclear waste, and half the book would be on the fact that nuclear power doesn't produce any greenhouse gases! Yet just thirty years ago, in 1977, a whole book had two pages mentioning the greenhouse problem.

I was in another project, called "Energy, the Next Twenty Years" (it's an embarrassing title now). We published a 600-page book in 1979. I recently looked in the index: there are ten scattered references to the greenhouse problem, adding up to less than ten pages out of a 600-page book. "Energy, the Next Twenty Years": in the next twenty years the greenhouse problem began to dominate all energy discussions. That's how new the subject is, and while I sympathize with people who get impatient about how little is being done to mitigate the problem, this is, nevertheless, still early in the era of concern with the greenhouse problem.

I mentioned that there is uncertainty about how much global warming may result from various levels of greenhouse gases, and then there's uncertainty about how those translate into climates. Then you have to translate those climate changes into the impacts on productivity, health, comfort, recreation, and everything else. There I find one of the greatest difficulties is imagining the kind of world we will be living in when these climate changes begin to be serious, fifty or one hundred years from now. How do you come to grips with the way life is going to change?

One possible way is to look back at how much life has changed in the last fifty or seventy-five years. Sixty years ago, we didn't have electronics, we didn't have nuclear power or nuclear medicine, we had no antibiotics, hardly any vaccines; there were no plastics—we had celluloid. If I ask myself: back in the 1920s, if people were predicting the kind of climate change that some people are predicting now for the next eighty years, how would people have reacted in the 1920s? I expect that they'd be concerned about what was going to happen to mud. In the 1920s, mud was a serious environmental problem. Automobile tires were about 2.5 inches in diameter, pumped up to 60 pounds per square inch, as hard as wood, and in mud, automobiles got no traction. My uncle Harry used to make money from cars that stalled in the muddy road that went by his farm; he'd take a team of horses and pull the cars out of the mud. You couldn't ride a bicycle in mud, and in those days most children walked to school, and mud was a problem. People would have wondered, "What's going to happen in the summertime when the climate changes?" maybe not realizing that by the end of the twentieth century, at least in the United States, everything would be paved solid—my children have never driven a car in mud, let alone my grandchildren.

I also ask myself, since the main impact of climate change is likely to be on agriculture, how would a ninety-year-old farmer respond if you asked him, "What has happened to farming in the eighty years or so that you've been working on a farm? What are the most dramatic changes?" He would likely say, "Well, the disappearance of the horse, the disappearance of kerosene lamps, the telephone; my car has a heater now." I'd say, "What about climate change?" and he'd say, "Well, it's hard to know what the impact on farm productivity has been because we have different crops now. We never had soybeans in the 1920s; now we grow soybeans. We have hybrid corn, we have insecticides and herbicides and things of that sort, artificial fertilizer, and everything is so changed, I don't know what to attribute to any climate change."

So much has changed in the last eighty years that it's important to recognize how much may change in the next eighty years. It's important not to imagine climate change being superimposed simply on life as we know it now.

We can be sure of a few things. One is that, in a developed country like France or the United States, not much economic production for the market is affected by climate. In my country, you can assemble automobiles in any state. We don't assemble them in Alaska, but that's because of distance rather than climate; you can do open-heart surgery in any state; you can do banking and insurance—well or badly—in any state of the Union; you can produce pharmaceuticals in any state; you can do television broadcasting in any state. Except for agriculture—farming, forestry, fisheries—there's not much that *is* affected by climate. In a country like France or the United States, despite the remarkable political power of farmers, farming is a small part of the economy. In the United States, farming is less than 2.5 percent of our GDP. Some may say, "But it's an important part: without farming you can't have food." The way to look at it is if the cost of producing food in a country like the United States would double or triple or quadruple over the next eighty years, it would probably mean that per capita income would double, not by the year 2060, but by the year 2063 or 2064. The impact of climate on GDP per capita in the developed countries is not likely to be large.

The situation is more acute in developing countries, where as much as half the population can depend on agriculture, much of it subsistence agriculture. So developing countries are vulnerable to climate change, not that all climate change will necessarily be adverse, but they are vulnerable in a way that Americans, Canadians, French, Australians, or Israelis are not vulnerable. So the countries that are least able to do anything about greenhouse gases are the countries that have the most at stake, the most to lose from any drastic climate change.

There's a lot of concern about the effect of climate change on health. For reasons that I don't understand, most tropical vector-borne diseases, whether malaria or schistosomiasis or river blindness or whatever it may be, become more virulent when the temperature goes up, and most of those diseases borne by mosquitoes, fleas, flies, gnats, and snails will spread territorially as the tropics expand with global warming. So the health impact has to be considered.

But again we want to recognize that things are likely to change over the next fifty years. A good example is Singapore and Malaysia. Singapore is separated from Malaysia by 1 kilometer of seawater—they have identical climates. Half a century ago, Singapore and Malaysia were part of the same nation. Since then, Singapore has developed to become about the highest-standard-of-living country in the world. Malaysia has also developed, but nothing like Singapore. There is no malaria to speak of in Singapore; there's lots of malaria in Malaysia. Now that's partly because Singapore is a rich, compact state that is able to eradicate mosquitoes. If a Singaporean gets malaria, it's probably because he or she spent the weekend in Malaysia and got bitten by a mosquito and came back having contracted malaria. But the Singaporean is healthy to begin with. Malaria kills a million people every year, but it doesn't kill healthy people; it kills people who are undernourished or already suffering some debilitating disease. The Singaporean who gets malaria in Malaysia gets excellent medical treatment. It's a serious illness, but it's not life threatening.

The same with measles: measles kills a million children every year. When I was a boy, everybody got measles. You stayed away from school for a week or ten days; it wasn't a terribly uncomfortable disease. How then does measles kill a million children every year? It's killing children who are weak to begin with, undernourished, sometimes hungry, but often just lacking the nutrients that would make them healthy.

If in the developing world, in the course of the next fifty years, we can reduce the incidence of childhood hunger and malnutrition, the impact of climate change on health, on longevity in the developing world, will be hugely attenuated. I mention all this to suggest that the best defense against climate change for the vulnerable parts of the world—meaning the so-called developing countries—is going to be their own development, so they become less dependent on subsistence agriculture and less vulnerable to the various kinds of diseases that can afflict them if they are undernourished to begin with.

That leads me to a somewhat pessimistic conclusion about the political response to the threat of climate change. It is going to be hard to get people in my country, the United States, to take climate change seriously, if they are made to believe that the vulnerable parts of the world are someplace else. Al Gore probably exaggerated the threat to people in my country, and I don't know how to get them to take it seriously unless we *do* exaggerate the threat to them, or else persuade them that peace and security in the long run depend on the successful development of the parts of the world that are still poor and vulnerable to climate change.

I want to talk a little about mobilizing the nations of the world to do something about global warming. I know of no peacetime historical precedent for the kind of international cooperation that is going to be required to deal with climate change. In wartime, the Soviet Union, the United Kingdom, and the United States energized themselves to defeat the Axis; nobody had to enforce the military commitments that were undertaken by the British, the Americans, and the Soviets. I can't think of anything in peacetime of the *magnitude* of the cooperation that is going to be required among the main nations of the world.

Two things are going to make any enforceable system difficult. One: I don't see any possibility that we could agree on what is the limit on the concentration of greenhouse gases in the atmosphere. Getting emissions down is a short-range objective. Eventually the problem will be to stabilize the concentration in the atmosphere, which means getting emissions down to where they're not adding to the concentration. The uncertainties are still at least as large as a factor of three, so to decide what the target is for the end of the twenty-first century is currently not feasible.

I also don't see any chance that we can have *enforceable* national limits on greenhouse gas emissions. People talk about "binding commitments": well, commitments are supposed to be binding, that's what the word means, but who is going to enforce the United States' living within a certain quota? Who's going to enforce it on Guatemala, or Namibia? Nobody is going to propose military force; I doubt anybody's going to threaten economic boycott. We are talking about what is inherently a voluntary system.

I was impressed a few years ago when both France and Germany were about to violate the fiscal provisions of the European Union. No nation was supposed to incur budget deficits in excess of 3 percent of GDP for three years in a row. In 2004, France and Germany were both just about to complete their third year in a row with budget deficits in excess of 3 percent of GDP—and nobody expected anything to happen, and nothing did happen. It's hard to imagine a greenhouse regime that would be tighter, more binding, more serious than the European Union, and if you can't enforce a primary rule on the members of the European Union, I don't see how you're going to enforce carbon quotas on major countries.

My only historical example of cooperation on the scale that is likely to be required is the North Atlantic Treaty. Beginning in 1951, fifteen nations negotiated over what they would do to build up defense forces against the possibility of attack from the East. Those nations did incur major commitments about drafting young men into military service and finding the budgetary means of equipping them, maybe letting them be stationed in Germany, finding real estate for maneuvers, for barracks, for military pipelines, and things of that sort. There was no enforcement. The North Atlantic Treaty Organization (NATO) nations simply incurred serious commitments and for the most part lived up to them—at great expense. But as I said, there was no enforcement mechanism. They met their commitments because self-respecting nations incur commitments, and as best they can, they meet those commitments.

So I propose that the way to mobilize the world in relation to the greenhouse problem is first to get the United States on board, and that is almost certain to happen now. Then the developed countries, maybe for convenience the Organization for Economic Cooperation and Development (OECD), could fashion themselves a little on the North Atlantic Treaty and negotiate what they will do. Here it is important to notice that in NATO, the commitments were not to results but to what was *actually to be done.* In Kyoto, the commitments were to the *results* that were to be anticipated by 2010 or 2012. Most nations that committed themselves at Kyoto didn't have any good idea of what would be required in order to meet those commitments. If there is to be a workable "successor" to Kyoto, it is important that the commitments to be undertaken by the major nations should be commitments to what they will *do,* not to what the results will be fifteen, twenty, or twenty-five years down the road.

One of the first things to do is research and development. Right now, in both the United States and Western Europe, energy research and development is much lower than it was in the mid-1970s. We took it seriously in the 1970s; apparently we don't take it that seriously now.

What kind of research and development? Well, one kind—Marty Weitzman mentioned the possibility—concerns capturing and sequestering carbon dioxide. It's been known for fifty years that you can pump carbon dioxide, as may be produced at an electric power plant, down into an oil well. It's been used that way to help extract oil from wells that have been substantially depleted. So the technology of getting carbon dioxide underground has been around for forty or fifty years, but it wasn't in anybody's interest to find out what happened to the carbon dioxide ten, twenty, thirty, or forty years later. So there's been no research and development on where it may stay once you put it there, or how you have to seal it to keep it from leaking out. We should engage in research and development on what kind of sequestration will work, keeping in mind that if you put it underground and it leaks out within fifty or seventy-five years, you haven't accomplished anything.

The Chinese are building coal-fired electric power plants at the rate of more than one per week. It's crucially important, if sequestration is going to be workable, to find out where to locate those power plants. It's cheaper to transport electricity than to transport carbon dioxide. So you want to build the plants close to where you can sequester the carbon dioxide, and the sooner you find out where that can be done, the sooner you'll know where these fifty or seventy-five plants per year that are being built should be located.

This is going to require a lot of geological exploration and experimentation and that should be getting attention in the United States, China, Russia, Germany, and anywhere that they're heavily dependent on coal. That's the kind of thing on which so little is being done. The United States had a pilot plant for sequestering carbon dioxide, and it was cancelled just a few months ago, and nobody's given a good explanation of why it was cancelled—whether it was because there is a superior technology coming up soon, or whether it was becoming too expensive, or whether there just wasn't enough interest.

Let me just mention now what some of the most serious problems may be from climate change. Marty Weitzman mentioned one: the possibility that the average global surface atmospheric temperature might not stay within the range of 3, 4, or 5 degrees Celsius, but that what he called the fat tail of the distribution might lead to an increase of 10, 12, or 15 degrees. He didn't explain the mechanism, but he did mention methane. He said there are enormous methane deposits all around the world on the continental shelves at the bottom of the oceans, but essentially the shallow oceans that are available to China, Mexico, the United States, and almost anywhere in the world, and nobody knows what might cause those methane deposits to be released.

Methane is a potent greenhouse gas. It doesn't have a long residence time in the atmosphere, ten or fifteen years, but if you put enough of it up there, it could aggravate the greenhouse problem. Right now, it looks as if there may be lots of methane trapped underneath the permafrost in Siberia, Canada, and Alaska, and some of that permafrost is beginning to soften and melt. Carbon dioxide is already being emitted from the melting permafrost, but the worry is that there may be lots of methane underneath that has been trapped in by the frozen surface, and when the surface becomes unfrozen, the methane may leak out. The problem there is that you may get a multiplier effect because the more methane leaks out, the more temperature will rise, and therefore more methane released. That's the kind of mechanism behind Marty Weitzman's worry.

The second thing to worry about is a body of ice called the West Antarctic Ice Sheet, which is what they call "grounded ice"—it's resting on the bottom of the ocean, but it rises a kilometer or two above the ocean level. When the Arctic floating ice melts, it does nothing to sea level, but this so-called grounded ice is like an iceberg so big that it's resting on the bottom, and if it should glaciate into the ocean or break loose from where it is and collapse into the ocean, the estimate is that there's about 6 meters' worth of rise in ocean level in that particular body of ice. When I was on a committee of the National Academy of Sciences in the early 1980s, we were worried about that, and we got glaciologists to look into the question of how likely it is that the West Antarctic Ice Sheet will collapse and raise the ocean level. We were assured that it could happen, but not for two, three, or four hundred years, and it would probably be gradual, and we would have plenty of warning.

By a strange coincidence, people got interested in Arctic and Antarctic ice just about the same time that we began to have satellite reconnaissance, and all of a sudden we can now measure the rate of glaciers' flow by the meter, not the kilometer. There's a lot of worry now that warming of the water in Antarctica may lead this West Antarctic Ice Sheet to break loose, and 6 meters of ocean level is a lot: Copenhagen is underwater; Stockholm is underwater; Los Angeles is underwater; the president has to go by boat to the Capitol.

You can save a lot of real estate with dikes, the way the Dutch have been doing for several hundred years, but you can't save a country like Bangladesh, where tens of millions of people live within 20 feet of sea level. If you could build enough dikes to save Bangladesh from the ocean, they would all drown in fresh water because you can't get rivers to flow up and over the dike. If you've ever been in Rotterdam, the dikes are against the rivers, not against the ocean, because the rivers are above or at sea level, and Rotterdam is below sea level. So this is one of the things to worry about.

A final word: Marty mentioned geo-engineering. Geo-engineering essentially means, as it's come to be used in the language of greenhouse problems, somehow reflecting away part of the sunlight, increasing what is known as the earth's albedo, the reflectivity of either the surface or the cloud levels or something of the sort. As Marty mentioned, it's been known that volcanoes that emit sulfur have a significant cooling effect on the earth. He mentioned Mount Pinatubo, which erupted in the Philippines in the early 1990s, and it had a measurable effect on air and ocean temperatures because it had a lot of sulfur in it.

So the proposal is, if we are changing the earth's climate by putting something in the atmosphere that impedes outgoing radiation, why can't we offset that by putting something in the atmosphere that intercepts *incoming* radiation? Now, sulfur is not considered a healthful thing to put in the air. On the other hand, the estimates are that if you could put appropriate "sulfur aerosols," as they are called, into the *stratosphere,* it would tend to stay up there for a year or so. Most of the sulfur that goes into the atmosphere in China or Germany or the United States has a residence time of about a week or ten days, and then it comes down. If you get something in the stratosphere that would last for a year or so, it would take a small amount of sulfur compared to the amount that is currently being put into the atmosphere all around the earth.

Undoubtedly, if we had fifty years to experiment and get ready, we could find something more benign than sulfur to put in the stratosphere. When I used to talk about this fifteen years ago, half the audience thought I was crazy and half the audience thought I was dangerous. But this is beginning to come out of the closet. There's going to be a large conference in Copenhagen in March, in which geo-engineering will figure heavily on the agenda, and this is beginning to get attention; we're hearing more and more about it. This is perhaps to be thought of as a last resort, in case over the next fifty years we find that doing anything to mitigate climate change has become politically impossible. But even as a last resort, the sooner there is some experimentation to find out how to perform this feat of geo-engineering, where to do it over the earth, and what to put in the stratosphere, the sooner we'll know whether this is likely to be a safeguard.

It has an acute disadvantage: one reason why people concerned with the environment don't like to talk about geo-engineering is that if you think this is going to be a cheap and easy solution to the greenhouse problem, you lose interest in developing the technology to reduce emissions of carbon dioxide. That's a dilemma. I don't know how to face it. But if it does turn out that geo-engineering becomes attractive, it will be helpful to have some experimentation to find out just what are likely to be the benefits of this kind of geo-engineering and what may be some of the dangers.

One of the dangers is that this looks to be such a *cheap* way of offsetting global warming that it might be within the economic capability of the United States alone or of China alone or of some state alone to engage in this, and that might mean that it becomes a matter of international conflict. But the sooner we discover whether this is going to be feasible and what some of the disadvantages may be to watch out for, the better we'll be. To give you some notion of magnitude, the amount of incoming sunlight that would have to be reflected away to offset a doubling of the concentration of greenhouse gases is around 1.5 percent of the incoming sunlight. Nobody bathing on the shores of the Mediterranean would notice the difference, and probably most astronomers with their telescopes would hardly notice the difference. It's a tiny fraction of the incoming sunlight.

Anyway, you will be hearing more about that in the years to come.

NOTE

1. Weitzman 2009.

REFERENCES

Ford Foundation, Resources for the Future. 1979. *Energy: The Next Twenty Years.* Cambridge, MA: Ballinger Publishing Company.

Nuclear Energy Policy Study Group. 1977. *Nuclear Power: Issues and Choices.* Cambridge, MA: Ballinger Publishing Company.

Weitzman, Martin L. 2009. "Some Basic Economics of Extreme Climate Change." Harvard University. http://www.economics.harvard.edu/faculty/weitzman/files/Cournot%2528Weitzman%2529.pdf.

Chapter 7

The Ethics and Politics of Preventing Political Violence

Cheryl Koopman and Bridgette Portman

Conversation

Bridgette Portman (BP): I'd like to know a bit about your background and how you became interested in psychology.

Cheryl Koopman (CK): My mother was psychologically minded. She loved to analyze politicians and people in general, so I grew up interested in understanding how people were thinking and what made them the way they were. I'm from a huge family, so I've always been embedded within a very large social network. When I was just finishing college, one of my brothers was severely burned in a fire. That got me very interested in the field of trauma in psychology because it was not only extremely hard for him, but it was extremely hard for my whole family to have a twelve-year-old child who was severely burned and lived apart from us while he was healing, and then upon returning, had to cope with others' reactions to seeing his burn scars. This wrenching experience made me very interested in understanding people's reactions to horrible events and situations.

I have been political since a very young age. I actually received a letter from President Kennedy in 1963. I wrote a letter after the Cuban Missile Crisis, telling him that I loved how he handled the crisis. I was only twelve years old. I received a letter from the White House stating that the president really enjoyed reading my letter. I think he might have because it was probably a funny and very gung-ho letter, so his staff may have given it to him to amuse him. I grew up interested in the war in Vietnam and civil rights, and I was an early feminist. I went to Berkeley starting in 1968, and I thought that the whole world was on fire with revolution and was kind of shocked a few years later when I realized a lot of the things that I thought had fundamentally changed still hadn't changed throughout the rest of the world. I was very actively involved in the demonstrations against the war in Vietnam and Cambodia, and I was involved later in the movement to overcome racism, the environmental movement, and the women's movement. Thus, I was interested in politics at the same time I was interested in psychology. As an undergraduate, I knew of other young people coming back in body bags. One of my neighbors died in the Vietnam War. He was just a year older than I was, a kid who had everything going for him; he was captain of the football team and very popular. So it was very real to me that political violence was affecting people's lives. I remember standing at Sather Gate on the UC Berkeley campus when I asked my philosophy teaching assistant whether the war in Vietnam was right or wrong. As he began to discuss why he thought it was wrong, a crowd of about a hundred and fifty people gathered around him with many loudly expressing their agreement or disagreement. It was a dramatic turning point in my life because it was the first time I'd heard the arguments against the war in Vietnam.

BP: That must have been an exciting place to be at that time. Where did you go after Berkeley?

CK: I went to UCLA and was involved with the Committee against Racism. We had students at UCLA from Iran who received phone calls in the middle of the night telling them that a family member had become seriously ill. They immediately travelled home by plane and were met by SAVAK, the secret police, and were never heard from again. These were students who were trying to organize against what was happening in Iran under the Shah, who was being supported by the United States. So I was aware that my own fellow students were being tortured and imprisoned and killed. I started a PhD program and then moved to the University of Virginia and earned my PhD there in educational psychology and program evaluation. I did a double major, and sociology of education as a minor, and psychology as a minor—even then I was interested in a lot of different things. After that I did eight years of post-doctoral work in three different fields: psychiatry, psychology, and political science. That training was at Harvard and Columbia.

BP: What do you see as the most serious instance of political violence going on today?

CK: I think Pakistan is a really dangerous place. It seems like it's in the news almost every day because of the violence that's going on there. There's great social instability, and it has nuclear weapons and a terrible conflict with India and many factions that see the United States as an enemy. There's so much violence right now in some of the African countries, like Sudan and Somalia. Some of the violence is not so much due to factions that are fighting a war, but to government policies that are not fairly distributing the wealth and are leading to people being threatened with starvation. We give an awful lot of our foreign aid, including food aid, to countries that actually don't need it. We do it for international security reasons. Egypt and Israel receive a lot of our foreign aid, and they're actually pretty well off compared to many other countries. The Palestinian conflict with the Israelis is extensive and highly threatening to the security of the world. Fortunately, there are many people inside of Israel who have great ideas and values and are working on both sides of the conflict to try to help resolve this conflict.

There's a couple of psychologists that I especially admire who have been working in Rwanda to try to help mend relationships between the Hutus and the Tutsis. They're American based: Ervin Staub, who is a Holocaust survivor, and his collaborator and partner Laurie Pearlman. They took methods that had already been developed in the West and brought them to people on both sides of the conflict, and asked them to critique the ideas and see whether there was anything applicable to their situation. I have tremendous admiration for their work.

BP: What do you think about the Obama administration, in terms of having the political will to try to intervene in some of these places or resolve some of these conflicts?

CK: I'm hopeful. I think our new president is really smart. I like the fact that he's consciously trying to overcome groupthink. But I also think it's hard to change policies from the course that they've been on. Obama has a dilemma: he wants to have people in his administration with political experience, but the downside of that is that they've already contributed to creating some of the problems that we now face, and they have mindsets that may not have changed sufficiently to deal with the new realities. Those people are very bright, but they also contributed to the mess, such as this economic downturn we're in, for example.

I'm partly hopeful because the situation is so bad, because I think a lot of times people are willing to make changes when situations get really bad. But it's amazing that people don't see the problems as they are developing. We have made so many short-sighted decisions where smart people should know better. Carter didn't see the problems that he could face with supporting the Shah in Iran. He toasted him on New Year's Eve before the Shah was overthrown. He supported bringing the Shah into the United States for cancer treatment. I attended a conference in which the former ambassador from Iran to the United States said that he personally warned the then Secretary of State, Cyrus Vance, that if the Shah came into the United States for medical treatment, there was a great chance that hostages would be taken at the American embassy in Tehran. He spelled it out and he was ignored. The Shah could have instead gone to Mexico and received the same medical treatment. It was symbolically so upsetting to the Iranians to see America's support for the Shah, especially because so many families were directly affected by the actions of the secret police. The Shah had also implemented some progressive policies, but the torture and other human rights violations were too horrible.

Elites didn't forsee the Berlin Wall falling until it was actually happening. Just before the Soviet Union collapsed, I did a fellowship in a research institute headed by Zbigniew Brzezinski, the former national security advisor. We had closed meetings with people in positions of power that pertained to national security issues—the president of the World Bank, the CEO of Mitre Corporation, people from the State Department and the Department of Defense. Brzezinski would always ask them at the end of the talk, "So what does the future hold?" And virtually every single speaker described a continuation of the directions inherent in the status quo. We were right on the verge of having the Soviet Union collapse, we were right on the verge of having Apartheid end in South Africa, and none of that was seen. So I think that to enable leaders like Obama to be more successful, it's really important to break their mindsets. That's why I feel that political psychologists can be helpful in encouraging them to think of the possibilities that are beyond the trajectories that they have been on.

BP: I wanted to ask about the broader ethical dimension that underlies your chapter, and that's the idea that we have an ethical duty to intervene to prevent harm. I wonder if you could talk about that a bit. For example, someone might argue that if we're not causing the violence, it's none of our business. Why do we need to try to intervene?

CK: Well, what are behind ethics are values. To maximize values often involves taking action. I agree with those who say that the utmost moral values have to do with social justice, promoting individual freedom for others as well as oneself, and reducing suffering. I also like the idea that promoting beauty is one of the moral values. So those are four ultimate moral values that enhance being alive and being human and provide a sense of meaning, and also promote quality of life and the survival of the human race. Erik Erikson's framework for what is a self-actualized, mature person includes generativity, looking out for future generations. I think it's part of mental health for adults as we age that we look out for other people's well-being. It's part of developing our full humanity. Also it's essential to the well-being of society that people do not just look out for themselves.

To internalize these values in the first place, I think it really matters that children are brought up in a context where they're exposed to these values. One of my earliest memories is of my parents asking my sister and me to pack up our old toys and clothes. We got in the car, and I remember going to a clearing next to a burned house where there was a family with their children standing by their car. My father handed the other father a stack of money, and we gave them blankets, our toys and clothes, and bags of groceries that my parents had just bought. The family seemed really happy to receive what we gave them, and I felt so proud that we were helping them. It was the first memory I have of doing something for other people and feeling really good about it. In retrospect, I realize that that was a wonderful gift our parents gave to my sister and myself.

BP: George W. Bush justified the Iraq War partly on a humanitarian basis, saying we had to free the citizens from Saddam Hussein. Some people argue that that actually caused more harm than it prevented. How do we decide when to intervene, and who makes that decision?

CK: That's a great question. I am influenced in my answer by Ernest Becker, who wrote *The Denial of Death* (1973) and *Escape from Evil* (1975). I believe that humans are attracted to transcendent concepts that can sound very benign and spiritual but at the same time can cause tremendous human suffering. The Nazis were no exception to that. They had a view that they considered moral, about destroying Jews and other groups of people in order to make society better, and they justified it on moral grounds. So they were the ultimate example of doing a lot of evil in the name of what they were trying to say was good. I'm also influenced by Rollo May, an existentialist, who wrote *Love and Will* (1969) and *Power and Innocence* (1972). He argued that we should try to hold ourselves responsible for the actual consequences of our behavior, rather than letting ourselves off the hook by saying our intentions were good. It has to do partly with our ability to hide things from ourselves. We have a preconscious awareness of the possible consequences of things, but if we just let ourselves off the hook by saying, "Oh, my intentions are good," and, "I know the answer to this, and I don't have to consider any other possibilities," then we are very susceptible to doing harm in the name of doing good.

One of the things I like about Obama is that he's consciously trying to surround himself with people who have differing points of view. I'm amazed that he chose Hillary Clinton as his secretary of state. I don't think he did it for purely cynical reasons; I think he really did it out of a deep sense that it would ultimately produce the most good, even though they had engaged in head-to-head conflict during the electoral campaign. So one of the ways to transcend this problem of trying to do good but ending up doing harm is to structure the process of policy-making in a way that criticism is not only elicited, but is actually listened to and considered. Another part of the answer is to set up indicators of what outcomes you expect a policy to lead to by certain dates, and if a policy is falling short of meeting those indicators, then that feedback needs to be addressed by changing the policy accordingly. In addition to that, there's a concept in program evaluation and policy evaluation that needs to be part of the process across the board, which is to look for the unanticipated consequences of one's actions. I think we're doing that right now, to some extent, by considering the unanticipated consequences of failed deregulation policies. That's more cognitively demanding; it's harder to do. But if you try, you can do it. Building these kinds of checks into our process will improve our decision making.

BP: In terms of preventing political violence, do you see a role for international organizations like the UN?

CK: I don't think we're going to survive if we don't have international organizations play a bigger role. We're not yet really dealing with global warming the way we should be. It's going to dominate our political agenda at some point in part because of the dangers of rising oceans. Bangkok, with its population in the millions, is one of a number of cities in danger because of being located just above sea level. I think within this next century, if we don't make major changes, it will be a question of the survival of human beings on the planet. But I think human beings will come through. Sometimes things just have to get really scary, and then we pay attention to them and make changes.

I'm also hopeful about another aspect of what Ernest Becker said in *Denial of Death.* He said that human beings have a need for transcendent concepts that they can identify with, whether religious or spiritual or just some kind of existential meaning that is inspiring. I think he used the word "cosmic heroism." We not only need these concepts, like freedom and so on, but we also need inspiring stories about human beings, like the god myths for the Greeks, but something for us that's relevant to our time. At Berkeley in the late sixties, we had rock stars who were to some extent playing that role for us. I still really look up to people like Bono. And we have people like Angelina Jolie and Brad Pitt and others who are trying to promote good values, like helping people to rebuild their homes in New Orleans. Barack Obama is the ultimate celebrity right now. He has such a high approval rating in the United States. But part of it is not just him as a person but people's images of him. Part of the story of Barack Obama is his humble beginnings, his community organizing, growing up with a very poor mother who died when he was still young. Also, the fact that he lived in Hawaii and then Indonesia and Chicago—all these different cultures—means that he embodies diversity. He represents a person who integrates differences. I think that's part of his cosmic heroism for the world. I think we need leaders who embody these values.

BP: You've done a lot of work on people living with serious illnesses. How does traumatic stress due to illness compare to that due to political violence?

CK: Human beings have a lot in common despite our differences, and our overall responses to stress are basically the same across all cultures and across both political trauma and medical illness trauma. Having said that, there are some differences, such as that medical stress tends to be more future-oriented and viewed as originating from within. This is a relatively new field—most of the work on traumatic stress response has been done since the Vietnam veterans began to return home in large numbers. Consequently, most of the research on traumatic stress is only about four decades old, which isn't very long, compared to many other fields of inquiry.

As far as we know, when people encounter a stressful situation, how they perceive the threat, what they think could happen in the future, and also how they perceive the loss that's presented by that situation affect their responses. Somebody could have a diagnosis of breast cancer and not be that traumatized by it because she's dealing with a very excruciating divorce, which is actually more upsetting to her than is the breast cancer. To some extent, how people think about a situation is important in shaping their responses, and that's true with political violence as well. On the other hand, breast cancer is potentially life-threatening, and political violence is also potentially life-threatening, so the effects of stress are not just in people's minds. People's responses at the time of such a stressor, no matter whether it's illness or political trauma, often include hyperarousal, intrusive thoughts, avoidance, and dissociation, which is a trance state that people experience. Our own research, as well as that of others, suggests that people who were traumatized early in life may end up learning to dissociate as part of their stress response, and so they may have greater dissociative responses to subsequent life stresses.

BP: Are you working on anything currently?

CK: I am now connected with a group of people in San Jose who work with torture survivors. Their Center for Survivors of Torture conducted focus groups with recent immigrants from various countries where there has been a lot of torture—Cambodia, Vietnam, Iran, Iraq, some Sub-Saharan African countries, and Bosnia. They didn't recruit people specifically on the basis of torture, but the issue came up. It's estimated that somewhere between 5 percent and 35 percent of refugees in the United States have been tortured in their home countries. It's not usually part of our rhetoric about immigrants; when we talk about how to deal with, for example, Mexicans and Central Americans coming across the border, we hardly ever talk about the fact that so many of them were tortured in their countries of origin. It influences how they access services here. For example, when I was living in New York City, there were a number of people from Cambodia who lived in housing with broken windows. The landlords wouldn't replace the windows, and the people living there were afraid to say anything because they assumed if they said anything it would be dangerous. They didn't assume they had rights. Therefore, in this collaboration, we are analyzing the focus group data to learn more about what these immigrants from different ethnic groups want in terms of social services and try to make sense of how that might relate to where they came from politically.

BP: You're currently president of the ISPP [International Society of Political Psychology]. How has that experience been for you?

CK: It's a great honor. I really wanted to be president of ISPP; I love the organization. I feel very connected to a lot of people in it. When you're president of an organization like ISPP, you don't have that much power because it's very egalitarian and power is shared among a lot of leaders. You definitely have some discretion and influence, and it's an opportunity to try to achieve some things, but I hope that I can accomplish more after I'm president because a year goes by so quickly.

REFERENCES

Becker, Ernest. 1973. *The Denial of Death.* New York: Free Press.
———. 1975. *Escape from Evil.* New York: Free Press.
May, Rollo. 1969. *Love and Will.* New York: W. W. Norton & Co.
———. 1972. *Power and Innocence.* New York: W. W. Norton & Co.

Ethical and Scientific Considerations of Traumatic Stress in the International Context

Cheryl Koopman and Nicole Wernimont

Investigation

Ethical problems in the political context arise when highly disturbing circumstances and events—including human rights abuses—threaten the well-being and survival of individuals and communities alike. Unfortunately, such circumstances and events are all too common in the international context, indicating a critical need for greater knowledge of how to effectively prevent and respond to such occurrences. In this chapter, we will discuss: ethical considerations that arise in this context; traumatic stress responses to political violence and adversity; and constructs drawn from research on traumatic stress that may help to inform the alignment of practice in the international context with ethical principles.

Perpetrators of violence can be judged according to multiple ethical considerations. For example, in judging the US decision to go to war against Iraq, the possibilities of promoting human rights and social justice and reducing human suffering by going to war should be weighed against assessments of the adverse effects of this action. In addition, in the international context, the problem of ethical responsibility often arises in the context of considering third-party intervention. Accordingly, citizens from less directly affected nations or those nations' representatives, such as US leaders or the UN, reflect and possibly act upon their responsibility to intervene in a dispute between two or more involved groups, such as occurred in response to the Darfur genocide. In such situations, considerations of what is possible to achieve, as well as issues of national sovereignty, human rights, social well-being, and special interests, have conflicting implications, creating uncertainty about what degrees of engagement and disengagement comprise ethical action. Although we do not claim to resolve or even thoroughly address the interaction of these forces in this chapter, we do suggest that the body of psychological knowledge introduced here ought to have some power in verifying the need for political engagement and intervention when the cultural and material resources for such engagement exist.

In addition to the diagnostic and etiological contributions that will be discussed, it should be noted that psychological science has also collected a great deal of data about the influence of social psychological processes on decisions on whether to engage in or avoid acts of political violence. Psychology has added considerably to our understanding of the influences of shared conceptual and linguistic practices on public perception and policy formation related to international human rights abuse intervention. For example, Philip Zimbardo (2007) has reiterated the notion that dispositional analyses of so-called evil-doers detracts from the importance of systemic and situational analysis, which he argues is more capable of yielding effective intervention strategies. The factors influencing whether political violence occurs and whether parties act to prevent or stop it are not a major focus here; however, we acknowledge the importance of such factors as they must be considered in making ethical judgments.

A dispositional perspective comes from Glen Greenwald, in his book *A Tragic Legacy: How a Good vs. Evil Mentality Destroyed the Bush Presidency* (2007). He describes a hazardous situation in which post-September 11, 2001, decision making was misguided by overly rigid thinking and simplistic dispositional attributions for understanding national security threats. Greenwald argues that the George W. Bush presidency had unintended devastating effects on subgroups of Americans and on people in other countries, including Iraq, because of this president's tendency to construe all political problems along the purported axis of good versus evil. Such a mentality, Greenwald suggests, may increase the risk of engaging in inappropriate actions. For example, he suggests that because the worst assessment of the Iraqi government was assumed, there was inadequate scrutiny about the faulty nature of the evidence for the presence of weapons of mass destruction that initially was used to justify the war in Iraq.

Critiques concerning the ethical dimensions of political forms of engagement, such as Greenwald's analysis of President Bush's motives for going to war in Iraq, are supported by social psychological research such as that conducted by Philip Zimbardo (2007) and Ervin Staub (1999). On one hand, engaging in faulty action is ethically problematic; on the other hand, equally problematic concerns arise from disengagement from ethical duty, which is also informed by psychological science. Following the highly publicized case of Kitty Genovese in New York City, Bibb Latané and John Darley (1968) experimentally demonstrated the propensity of observers to fail to intervene on behalf of strangers in emergency situations. However, these and other researchers have also demonstrated the propensity of observers to effectively intervene in emergency situations when in the presence of peers or authority figures who themselves act on behalf of helping a stranger (Latané and Nida 1981). This effect ought to bear significantly in our analysis of ethical responsibility on behalf of those suffering in international crises as well. If people are suffering and bystanders in a position of greater strength fail to act in helpful ways, this is considered unethical to the extent that the opportunity is ignored that could result in alleviating and preventing further suffering. International examples of this concern have been widely discussed in response to acts of genocide, in which lives potentially could have been saved among Jews in the Nazi Holocaust (Miller 2004), Tutsis and Hutu moderates in the Rwandan civil war (Beardsley 2008), and African blacks currently living in the Darfur region of Sudan (Beardsley 2008). However, ethical considerations of how best to respond to such extreme levels of political violence are challenged by diverse viewpoints on how to do the most good while doing the least harm in such situations as Darfur (e.g., de Waal 2007).

The insights from social and behavioral science on this topic allude to the importance of effective leadership (Hamburg, George, and Ballentine 1999). To the extent that leaders can avoid the social and situational factors that inhibit effective interventions and instead cultivate factors that facilitate such interventions, they can make sound use of the gains enjoyed in our field over the past several decades. Misuse and ignorance of psychology and other disciplines undermine our collective responsibility to intervene on behalf of those whose basic rights to psychological well-being are violated for political gain. An example is the misuse of "the scientific methods" for conducting political interrogations and intimidation of prisoners, as alleged to have been taught by US and Israeli counter-insurgency specialists to the domestic security and intelligence agency (known as SAVAK) of the former regime of the Shah of Iran (Afshari 2002, 293).

Furthermore, conceptualizing the ethics of traumatic stress in the political context needs to encompass not only simple acts of political violence as sources of traumatic stress, but also political actions taken in the aftermath of political violence that further undermine human well-being and human rights. Political responses in the aftermath of traumatic situations can also lead to additional ethical costs when harm is further exacerbated by such political actions. This can occur due to a lack of consensus about how to best respond in the aftermath of political traumas such as the violence in the Darfur region of Sudan (e.g., see de Waal 2007). Worsening an already-horrible situation following instances of political violence can also be particularly malevolent when it is intentional, as argued by Naomi Klein (2007) in *Shock Doctrine: The Rise of Disaster Capitalism.* Klein argues that a number of US presidents and other world leaders have perpetrated tremendous harm by following the approach described by former economic advisor Milton Friedman, who recommended taking advantage of the vulnerabilities associated with the aftermath of disasters to make desired political changes such as government deregulation and privatizing public assets at times when populaces are too vulnerable to resist such changes. Political actions taken in the aftermath of disaster that take advantage of the survivors' vulnerabilities in order to enact political changes that they would otherwise resist have multiple ethical costs; not only do they often worsen human suffering, but also such actions intentionally politically disempower vulnerable populaces.

To promote lucidity in ethical analysis of traumatic stress in the political context, it is important to consider not only social psychological principles but also the scientific thinking about traumatic stress in order to gauge the risk and severity of harm resulting from political action/inaction. We turn now to providing an overview of types of traumatic stress and some of the major factors affecting the risk and severity of traumatic stress resulting from political violence and other political behavior. Based on the current status of scientific understanding of traumatic stress and of the limits of this knowledge, we will conclude this chapter with a discussion of individual and contextual factors that must be considered in examining political violence from an ethical perspective.

CURRENT UNDERSTANDING OF TRAUMATIC STRESS

Psychological stress has been defined by psychologists and psychiatrists in various but specific ways. Traumatic stress, not to be confused with *exposure* to traumatic events, refers to the psychological reactions to such exposure. Traumatic events of relevance to international studies include genocide, war, terrorism, political incarceration, torture, assassination, displacement, forced migration, large-scale natural disasters, famine, and pandemic illnesses (e.g., AIDS).

Posttraumatic stress disorder (PTSD), acute stress disorder (ASD), and peritraumatic stress are types of traumatic stress that have been identified by psychologists and psychiatrists. These differ temporally: PTSD symptoms must be present at least a month after a traumatic event, ASD symptoms must be present within four weeks after an event, and peritraumatic stress occurs at the time of the traumatic event. All three types of traumatic stress emphasize symptoms of anxiety/hyperarousal, the re-experiencing of memories of the traumatic event, and avoidance of reminders of the traumatic event (American Psychiatric Association 2000). A fourth type of symptom, dissociation, which is a trance-like state that will be discussed later in detail, is also linked to these three types of stress, although its centrality to these forms of stress is a major subject of debate in the field of traumatic stress (e.g., Ginzburg et al. 2006; Ginzburg et al. 2009).

POLITICAL VIOLENCE AS TRAUMATIC EXPOSURE

Traumatic stress in the international context is not experienced in a vacuum; rather it is experienced in reaction to events that are highly distressful. In the international realm, exposure to traumatic events often involves exposure to political violence. Political violence takes many forms, and political trauma is equally heterogeneous. The traumatizing effects of political violence often result from exposure to war-time events. Wars expose populations and groups to a variety of more specific sources of stress, such as combat exposure, torture, and forced migration. Other forms of political violence occurring with an international context are acts of terrorism, political execution, and assassination. It is important to note that the traumatizing effects of exposure to these forms of political violence are not limited to those who survive combat, torture, or terrorism, but also may affect those who perpetrate these forms of violence (Beckham, Feldman, and Kirby 1998). Furthermore, in times of political violence, the close relationship between perpetration and victimization that is often at play may be a critical distinction between this and other forms of trauma.

Similarly, the role of malevolence distinguishes political violence from other forms of traumatizing experience including exposure to violent accidents and natural disasters. Indeed, several contemporary researchers have isolated the variable of malevolent intent as an important factor in the unique psychological consequences of political trauma (Pratchett, Brown, and Bongar 2006). According to this research, malevolent intent is a uniquely pathogenic factor in political violence.

One compelling perspective suggests that because malevolent intent presents a particularly threatening assault on personal and communal resources, including belief in the possibility of justice, meaning, and benevolence, it has the capacity to shatter some of the most important systems of meaning around which personality and other fundamental cognitive schema are organized (Hobfoll, Dunahoo, and Monnier 1995). The loss of fundamental networks of meaning may, in turn, lead to increased rates of PTSD and more intractable posttraumatic symptoms. This link has been made tenable according to research presented by Kessler and associates (1995) who found a positive correlation between the presence of malevolent intent and the development of PTSD. In their study, survivors of malevolently intended forms of trauma, including rape, combat, and threat of assault with a weapon, had dramatically higher rates of PTSD than did survivors of trauma that lacked an element of malevolence.

The importance of political trauma as a distinct form of suffering has been consistently underscored by research demonstrating high prevalence rates of PTSD in populations exposed to these types of political violence. Lifetime prevalence rates for male Vietnam veterans, for example, have been estimated as high as 30.9 percent (Kulka et al. 1990), and rates for currently deployed US armed services personnel who have served successive deployments in Iraq are beginning to draw near to this high rate, with current estimates around 15 to 20 percent ([US] Army Medical Command, cited in Zoroya 2008). Similarly high rates of PTSD have been estimated among survivors of torture and mass violence in the war-torn countries of Algeria, Cambodia, Ethiopia, and Palestine. In Algerian war survivors, for example, rates of PTSD have been estimated as high as 37.4 percent (de Jong et al. 2001).

SPECIFIC FORMS OF POLITICAL VIOLENCE THAT LEAD TO TRAUMATIC STRESS

In the following sections, we provide a brief overview of several types of political violence and the corresponding areas of related research on traumatic stress in international studies: genocide, terrorism, torture, kidnappings, disappearances, assassinations, and refugees. This is a sampling of the current status of research, which suggests that although traumatic stress has received some attention in the international context, much more research is needed.

Genocide

Survivors of genocide have likely witnessed many forms of violence, experienced an enormous amount of fear for their own lives, and suffered many personal losses. Likewise, survivors of genocide have invariably fought to survive in environments where the entire community with which they identify has been subject to the same magnitude of suffering. Clearly, this form of political violence exerts a uniquely powerful psychological effect on survivors (see, for example, Weine et al. 1998).

Research investigating the psychological consequences of genocide has isolated several of the variables unique to genocide that are likely to result in the development of posttraumatic psychopathology. In their study of child survivors of the Rwandan genocide, for example, Schaal and Elbert (2006) found that 41 percent of those surveyed had witnessed the murder of one or more parent. Similarly disheartening, another study of Rwandan child survivors found that 90 percent of those interviewed thought they would die and that 15 percent of those who hid to survive did so under piles of dead bodies (Dyregrov et al. 2000). Similarly, 75.4 percent of a large sample of Rwandan adult survivors interviewed for another study were forced to leave their homes, 73 percent had lost a close family member to violence, and 70.9 percent experienced the loss or destruction of personal property (Pham, Weinstein, and Longman 2004). In each of these studies, high rates of posttraumatic stress were observed among the majority of child and adult survivors of the Rwandan genocide. Unfortunately, the trauma associated with genocide has long-term consequences. Perhaps in the case of genocide, more than in other forms of political violence, the intergenerational effects of trauma can be seen (Danieli 1998).

Terrorism

The threat of terrorism is psychologically distinct from other forms of political violence in several important ways. Terrorism is, in part, characterized by the protracted and indefinite stretches of uncertainty and fear experienced by those under threat of attack. Unbound by geographic location and unpredictable in nature, terrorism is a particularly powerful form of psychological warfare (Fremont 2004;

Pratchett, Brown, and Bongar 2006). Similar to other forms of political violence, however, the psychological effects of terrorism are often severe, long-lasting, and difficult to treat.

Prevalence rates of posttraumatic stress have been shown to increase substantially in populations following terrorist attacks. For example, 44 percent of Americans self-reported at least one posttraumatic symptom in the first week following 9/11 (Schuster et al. 2001), and 7.5 percent of Manhattan residents met full criteria for PTSD during the same time frame (Galea et al. 2002). Even more alarming, 36.7 percent of those in the World Trade Center at the time of the attacks met full criteria for PTSD. This rate is similar to that found among survivors of the Oklahoma City Bombing who were on-site during the attack (North et al. 1999).

Although some abatement of posttraumatic symptoms has been observed following these and other terrorist attacks, the effects of terrorism also endure for many survivors (Tucker et al. 2000). Children may be particularly vulnerable to intractable symptoms following terrorist attacks. Research has documented a prevalence rate of PTSD ranging from 28 to 50 percent among children exposed to terrorist activity, many of whom continue to experience new symptoms after the threat of terrorism has passed (Fremont 2004).

In fact, recent research has focused on the impact of terrorism on children specifically and has yielded rich information about the harmful and long-lasting damage caused to children exposed to acts of terrorism. In her summary of the extant literature, Fremont identifies ASD, PTSD, behavioral problems, developmental difficulties, and significant increases in separation anxiety and agoraphobia as the most frequent psychological sequelae found in affected populations of children. The symptom profile found in adult populations also frequently includes those symptoms associated with ASD, PTSD, depression, and anxiety. A related and significant finding has shown that children and adults who suffer from posttraumatic fear and anxiety have been observed over-generalizing the threat associated with terrorism to other, unrelated concerns (Fremont 2004). Over-generalization of this sort has been shown to have a deleterious effect on the individuals suffering from anxious symptoms as well as on the economies of societies affected by terrorism (Pratchett, Brown, and Bongar 2006).

Importantly, the role of the media has also been implicated in the development of posttraumatic symptoms. Populations indirectly affected by terrorist attacks that have witnessed such attacks primarily through television exposure have also been found to develop significant posttraumatic symptoms (Pratchett, Brown, and Bongar 2006).

Torture

Torture is one of the most inhumane and psychologically injurious forms of political violence. Torture includes the physical and psychological subjugation of one human being to another in which pain is inflicted for the purposes of political gain. The physical and psychological violence typical of torture can have long-lasting and complicated psychological effects (Gerrity, Keane, and Tuma 2001). Such effects are being increasingly well documented in survivor populations worldwide. Information yielded by this research ought to facilitate informed action in response to the continued sponsorship of torture in the world today.

Survivors of torture often suffer from a wider range and more severe psychiatric symptoms than survivors of other forms of political trauma. Research conducted by Kozaric-Kovacic et al. (1999), for example, compared the posttraumatic symptomatology of Croatian combat soldiers and prisoners of war. They found that prisoners of war, who had all been physically and psychologically tortured during the period of their detention, exhibited nearly twice the rate of PTSD as combat soldiers. Similarly, Silove et al. (2002) found that Tamil survivors of torture suffered from more posttraumatic symptoms than other survivors of war-related trauma. Mills et al. (2005) also found that torture has long-term mental health consequences distinct from those of forced migration in Tibetan refugee populations. Among the populations studied in their research, 78 percent of those tortured were suffering from symptoms of PTSD, 23 percent of whom met full criteria for diagnosis. Together these and other studies confirm that the psychological effects of torture are uniquely severe.

Increased rates and severity of posttraumatic stress symptoms among survivors of torture may reflect the compounded form of trauma experienced by victims of torture. Torture victims have often been kidnapped, have disappeared, or have otherwise been illegitimately detained and are often subjected to sensory disorientation and denied basic needs. They are also likely to be taken from politically volatile environments in which they may have already experienced the loss of loved ones, personal injury, or protracted fear and uncertainty.

The effect of exposure to multiple traumas has been well documented, and the role of sensitization is no less significant among those subjected to torture. Survivors of torture may experience decreased resilience in the face of subsequent trauma. Loutan et al. (1999) found that refugees with histories of torture were more likely to develop symptoms of posttraumatic stress and depression. This relationship was particularly salient regarding the prevalence of re-experiencing symptoms. Loutan et al. (1999) found that three times more refugee survivors of torture suffered from re-experiencing symptoms than their counterparts. Similarly, parental experiences of torture were highly correlated with the development of psychiatric symptoms in refugee children in another study (Silove 2002). The relationships between torture, refugee status, and mental health must be further investigated given that an estimated 5 to 35 percent of refugees in Western countries have been tortured (Baker 1992).

Particularly when reviewing the literature for ethical implications, it is important to note that psychological forms of torture have a deep, enduring, and debilitating effect. Even where severe physical methods of torture are endured, survivors rate psychological methods—including the complete withholding of information regarding the whereabouts and well-being of family members and friends, the provision of false information about these same concerns (Kozaric-Kovacic et al. 1999), the destruction of religious signs, the abuse of a friend or relative, and fear for one's own life (Mills et al. 2005)—as the most significant sources of stress.

Kidnapping, Disappearances, Assassinations

Kidnapping and disappearance are practiced in conflicts across the globe. The uncertainty and fear visited upon loved ones in the wake of these acts should not be underestimated. The kidnapping and/or disappearance of loved ones has been regarded as the most traumatic experience endured by survivors of political violence (Mills et al. 2005). In several conflicts during the past decades, children have been particularly susceptible to the effects of disappearances. In one study, children whose parents had disappeared exhibited more posttraumatic symptoms and more enduring symptoms than children of parents known to have been assassinated (Munczek and Tuber 1998). In another study, a significant number of disappearances and kidnappings were linked to the recruitment of Tamil children as soldiers for the Sri Lankan army during the 1980s (Somasundaram 2002).

The Situation of Refugees

Political refugees comprise a vast and heterogeneous group of 16 million people worldwide (United Nations Secretary General 2008). Collected under the common heading of their refugee status, this population includes people with unique political and personal histories. However, personal histories of traumatic exposure are not uncommon among refugee populations, and the psychological effects of such exposure often complicate the ability of refugees to adapt to their new environment (Loutan et al. 1999). Histories of traumatic experience among refugee populations are likely to include suffering related to the often volatile political environments of their home countries and might include the loss of close friends and relatives to war-related violence or to the experience of inadequate access to basic resources such as shelter, food, and water.

In fact, these types of traumatic experience have been well documented among refugee populations worldwide. Witnessing acts of violence in their home country was the most significant predictor of PTSD in one population of Sudanese refugees living in southern Sudan, where the prevalence rate of the disorder was estimated at 46 percent (Karunakara et al. 2004). Similarly, in a separate study of children in a population of Cuban refugees, witnessing violence was the factor that was the most highly correlated with poor mental health (Hjern, Angel, and Jeppson 1998). As demonstrated in the aforementioned research, the effects of posttraumatic stress endure beyond the traumatic experience itself. This phenomenon has been widely observed by many researchers and remains one of the most important signifiers of ethical responsibility in the realm of mental health and international policy.

As a result of previous exposure to political trauma and the ongoing stressors related to refugee status, refugee populations such as these are particularly vulnerable to suffering from symptoms of posttraumatic stress. Sensitization to traumatic experience—in which an individual's earlier exposure to stressors results in greater stress reactions to subsequent stressors—has been well documented in the literature. For example, in research by Sourander (1998), exposure to repeated traumatic stressors was associated with considerable impairment in children's ability to cope with subsequent stressors (Sourander 1998).

Perhaps in response to the accumulated effects of traumatic experience, which are complicated by the conditions of life as a refugee, prevalence rates for PTSD among refugee populations is often very high. For example, 57 percent of Cuban children previously detained in refugee camps were found to have moderate to severe PTSD (Rothe et al. 2002). Similarly, 46 percent of refugee children from Chile, Lebanon, Turkey, and Iran were found to have "poor mental health" four to six months after seeking asylum in Sweden, a prevalence rate three times that of children sampled in the non-refugee population (Hjern, Angel, and Jeppson 1998). Rates of PTSD were also very high, 36 percent, among refugees at Kakuma refugee camp in Kenya (Kamau et al. 2004). A somewhat lower range was identified by others who documented a PTSD prevalence rate of 15 to 47 percent in epidemiological studies of refugee and post-conflict populations internationally (Hynes and Cardozo 2000; Kamau et al. 2004; Modvig et al. 2000; Mollica et al. 1998, 1999).

Psychiatric symptoms most often observed in refugee populations include those associated with PTSD, depression, and anxiety, but may also include the very severe symptoms associated with schizophrenia, bipolar disorder, and neuropsychological disorders, including epilepsy (Kamau et al. 2004).

Problems associated with refugee status and relocation are thought to complicate the symptom picture and healing process of those affected by experiences of political violence. Such problems include marginalization, discrimination, unemployment, asylum status, poor parental mental health, the birth of new babies, lack of social connection, and the duration of uncertainty faced during requests for asylum (Hjern, Angel, and Jeppson 1998). Refugees are likely to have left behind significant support networks and other coping resources to which they might ordinarily turn in the aftermath of traumatic experience. This may be further complicated in situations where those left behind are themselves vulnerable to political violence—resulting in uncertainty about their whereabouts, health, and safety—or are in situations where loved ones have been killed. Resource depletion is an important aspect of the loss experienced by refugee populations who may become less and less able to cope with subsequent stressors.

This sobering picture of the stress faced by refugee populations also yields information about how to best respond to psychiatric illness in these populations.

Community clinics have been successful in some refugee camps (Kamau et al. 2004), and researchers have identified the importance of family cohesion, social support, and basic needs in the readjustment process of traumatized refugees. From even this dearth of information, it is apparent that there exists a reciprocal relationship between the mental health of refugee populations and the economic, political, and social well-being of the environments to which they migrate.

CONSEQUENCES OF TRAUMA

We will focus here on five outcomes linked to traumatic stress that may be of potential interest for better aligning practice to ethical principles in responding to international violence and adversity. These are (1) traumatic memories; (2) dissociative states of consciousness; (3) changes in worldview; (4) resource depletion; and (5) increased sensitization to stress in individuals and future generations.

Traumatic Memory

It is widely thought that traumatic memory differs qualitatively from other types of memory in aspects of content, temporality, and accessibility. These differences may be particularly salient in the memories of individuals who meet criteria for PTSD (Berntsten, Willert, and Rubin 2003).

Particularly within populations suffering from PTSD symptoms, the content of traumatic memories is likely to be more emotionally laden and sensorial than ordinary memories (Brewin 2007). This qualitatively unique feature of traumatic memory has been well demonstrated by Hellawell and Brewin (2002) in their study of the difference between flashbacks and other types of non-traumatic memory. In their study, participants narrated traumatic memories using more perceptual and sensorial words related to seeing, hearing, smelling, tasting, and other somatic sensations than for ordinary memories.

As has been observed in clinical settings for quite some time, however, traumatic memories are also often marked by a conspicuous lack of detail and cohesion (Tromp et al. 1995). Contrary to experimental work, which has shown a positive correlation between emotionally salient experience and the retention of detailed memories, traumatic memories are often fragmented, lack important information, and do not include details about the peripheral aspects of the traumatic experience (Butler and Spiegel 1997; Kuehn 1974).

An explanation of this ostensible contradiction has been forwarded by Butler and Spiegel (1997) who propose a dose-response relationship between traumatic experience and memory retention. In their formulation of this relationship, the arousal of negative affect results in heightened retention of detailed memory up to a point, after which it has a qualitatively distinct and disorganizing impact on memory. In their formulation, this threshold is marked by a state of psychological absorption into the central features of the traumatic event, the meaning of which overwhelms existing cognitive schema and prohibits coherent storage of explicit memory. In this state of absorption, they argue, the peripheral details of an event may not be processed, and even the most important central features may be relegated to storage as implicit memories.

As a result of this process, the experience of remembering traumatic memories may also be quite distinct from that of ordinary memories. Memories rich in sensory information, for example, may be overwhelming and experienced as if they are presently occurring. Such memories may take the form of intrusive or repetitive flashbacks or nightmares, as described by the diagnostic criteria for PTSD (APA 2000). Similarly, such memories may have been stored solely as implicit, procedural memories and may only be accessible in the form of motoric or behavioral re-experiencing (Butler and Spiegel 1997).

Dissociative States of Consciousness

Dissociative states of consciousness seem like a ripe area for application to understanding irrational decision making and behavior in the international context. Dissociation is a trance-like state involving a *lack* of integration of memory, perception, and/or identity (Spiegel and Cardeña 1991). Specific symptoms of dissociation have been integrated into the ASD diagnosis (APA 2000) and include derealization, depersonalization, lack of awareness of one's surroundings, emotional numbing, and amnesia for important details about the traumatic event. Little has been done to examine dissociative states in the political context; however, there has been interest in studying dissociative reactions to military stress in order to identify individuals at risk for experiencing health problems following traumatic exposure. For example, Dimoulas and her colleagues (2006) have found that among healthy females undergoing stressful survival training in the US Navy, those who experienced the most dissociation during and following the survival training were significantly more likely to experience post-stress health symptoms.

Another important area for research is to examine dissociative states in relation to irrational behavior in crisis decision making in international affairs. As an illustration, US President George W. Bush was videotaped staring off into space for several minutes immediately upon learning of the terrorist attacks on September 11, 2001. It is quite possible that he was experiencing a dissociative state, as he failed to react for a surprisingly long duration of time. The possible significance of such a reaction was well documented in our research on dissociation in survivors of a major firestorm in California, which found that people who reported a high level of dissociative experiences at the time of the firestorm, compared to those who did not, were more likely to engage in irrational behavior (for example, cross police barricades and try to get closer to the fire, especially if they were more exposed to the fire in the first place) (Koopman, Classen, and Spiegel 1996). Those who dissociated were also *less likely* than others during the firestorm to pack their most important possessions when evacuating from the fire, suggesting that they were less likely to engage in active coping. Thus, during international crises such as the September 11 attacks, it could be expected that at least some political leaders could experience dissociative states that temporarily affect their ability to make political decisions.

Changes in Worldview

Horowitz (1978) and other trauma researchers have noted that changes in worldviews often occur in response to traumatic events. Underlying their work is the notion that aspects of traumatic events (e.g., a breakdown in a sense of safety) are usually discrepant with pre-existing views of the world. Therefore, traumatic stress involves the processing of traumatic experiences, which often entails changes in these views. McCann and Pearlman (1990) have identified seven domains in which beliefs, assumptions, and expectations about the world often undergo modification following exposure to traumatic events. These seven domains correspond to fundamental psychological needs: frame of reference (super-ordinate beliefs, such as that the world is "meaningful"), safety, dependency/trust of others and self, power, esteem, intimacy, and independence. In an Internet sample of 1,657 adults (93 percent American), we found that *fewer* negative changes in such existential views in the immediate aftermath of the 9/11 terrorist attacks significantly predicted both greater well-being and less distress six months later (Butler et al. 2009). This suggests that psychological adjustment depends partly on maintenance of adaptive views of the world in the aftermath of societal traumatic events (see also Janoff-Bulman 1992).

As a natural leap from these findings, one wonders how much traumatic stress exposure affects subsequent political views by modifying views of the world. A commencement speech that the then national security advisor Condoleezza Rice gave at Stanford University in June 2002 suggests that her earlier life exposure to a traumatic event may have had long-term effects on her political views and policy preferences. In her speech, she said:

> Terrorism is meant to dehumanize and divide. Growing up in Birmingham, Alabama, I saw the "home-grown terrorism" of that era. The 1963 bombing of the 16th Street Baptist Church was meant to suck hope out of the future by showing that hope could be killed—child by child. My neighborhood friend, Denise McNair, was killed in that bombing, and though I didn't see it, I heard it a few blocks away. And it is a sound that I can still hear today. Those memories of the Birmingham bombings have flooded back to me since September 11. And, as I watched the conviction of the last conspirator in the church bombing last month, I realized now that it is an experience that I have overcome but will never forget. And so it will be for all of us, you and me, who experienced September 11. The story is

repeated—time and again—in the Middle East, in Latin America, in Africa—and it came home to America.

Such reflections are consistent with the possibility that political leaders' previous exposure to traumatic events—even ones that do not directly involve international issues—may have major effects in shaping their views of and subsequent responses to international events, such as the 9/11 attacks.

Resource Depletion

The impact of traumatic experience cannot be comprehensively understood solely through analysis at the individual level. Rather, comprehensive analysis must take into consideration the impact of traumatic experience on families, communities, and society at large. This is particularly true when considering the psychological effects of political trauma. Political violence affects large groups of people and is often intentionally aimed at exerting a widespread effect (Pratchett, Brown, and Bongar 2006). As a result, the aftermath of political violence is often characterized by a ripple of traumatic stress throughout the aforementioned ecological circles of society (Hobfoll, Dunahoo, and Monnier 1995).

This ripple may exert its impact most forcefully through chains of resource depletion, known as "resource caravans," in which loss of one resource, such as physical health, is likely to lead to loss of additional resources, such as optimism and social support (Hobfoll 2001). Conservation of Resource theory (Freedy et al. 1992; Hobfoll 2001; Hobfoll, Dunahoo, and Monnier 1995) articulates the significance of resource depletion as a primary mechanism of traumatic stress. According to their formulation, resource loss is a uniform consequence of traumatic stress that occurs along four domains. Object resource depletion may include the loss of housing, personal transportation vehicles, or clothing, and condition resource depletion may include the loss of employment status, involvement in organizations, or leadership roles. Personal resource depletion may include the loss of hope, self-esteem, or a vision of the future, while energy resource depletion may include the loss of financial resources, time, or motivation. Resource loss of these kinds has been well documented and shown to contribute to the effects of traumatic experience at all levels of society. Exacerbating the psychological effects of traumatic experience, the loss of resources in each of these domains can undermine individual and collective efforts to cope with traumatic stress. A reduced ability to cope can lead to increased stress and further loss of resources. Together with mounting stress and heightened requirements for effective coping, this accumulation of loss can result in a spiral of increased debilitation (Hobfoll 2001; Hobfoll, Dunahoo, and Monnier 1995).

Resource loss may be a particularly relevant factor in conceptualizing the perpetuation of political violence. It has been argued that survivors of political violence suffer the loss of resources in several of the domains described, but perhaps most importantly in the domain of personal resources. Judith Herman (1992), among others, has argued that the overwhelming impact of traumatic violence exerts a disorganizing effect on personal meaning and previously coherent personal narratives. In an attempt to rectify this loss and take action within a newly formed narrative that incorporates the traumatic effect of political violence, some individuals may engage in acts of counter-violence (Speckhard and Akmedova 2004). Accordingly, understanding the significance of this type of loss is an important goal for continued research in this area.

Increased Sensitization to Stress

In general, the research suggests that the more traumatic the events to which people are exposed, the more likely they are to develop PTSD. This occurs through sensitization, which is a biological as well as psychological process. As Charney and colleagues articulated, "Sensitization generally refers to the increase in behavioral or physiological responsiveness that occurs following repeated exposure to a stimulus" (Charney et al. 1995, 280). Sensitization to traumatic stimuli has been documented in a number of studies with different populations, including Israeli soldiers (Solomon 1989), Vietnam veterans (Bremner et al. 1993; King et al. 1998), firestorm survivors (Koopman, Classen, and Spiegel 1994), Holocaust survivors (Yehuda et al. 1995), and rape survivors (Resnick et al. 1995). Such sensitization is thought to occur as a result of a number of biological abnormalities associated with PTSD, such as reduced hippocampal volume that allows less reasoning to occur under stress and an increased startle response (Orr and Pitman 1999; Shalev 1999). These changes occur as a cumulative effect of stressors on the physiological response systems and have been termed "allostatic load." Repeated stress-response activation have been shown to be associated with neuroendocrine dysregulation and adverse health consequences.

It is worth noting here that, in many cases, the effects of political trauma also extend beyond the lives of those directly exposed to political violence and can disrupt the lives of generations thereafter (Danieli 1998). The literature on the topic of intergenerational trauma has been uniquely molded by the controversial perspectives of researchers and survivors alike; however, it persuasively suggests that exposure to the effects of political trauma can be conferred between generations and that the impact of posttraumatic symptoms can be sustained through this passage (Auerhahn and Laub 1998). While a considerable portion of this literature has focused on the multigenerational effects of trauma in families affected by the Nazi Holocaust, important work has also explored the intergenerational effects of other forms of political violence (Danieli 1998). This and other instances of genocide, World War II, the Vietnam War, and the enduring effects of repressive governments have all contributed to our understanding of intergenerational trauma. In the aftermath of each of these theaters of political violence, researchers and survivors have documented posttraumatic stress symptoms as they have been passed from parents to their children and through the shared rehabilitative processes of entire communities.

FUTURE RESEARCH EXAMINING TRAUMA IN THE INTERNATIONAL CONTEXT

Hopefully this brief discussion has highlighted possibilities for drawing from psychological perspectives on traumatic stress to add to our understanding of issues in the international context that have ethical implications. To better promote the ethical principles of reducing suffering and promoting social justice and human rights, we need to better understand the effects of international violence on broader indicators of quality of life than addressed here—not limited to more individual-level factors such as traumatic stress, but including effects on more societal levels, such as family cohesion, economic well-being, public health, and political engagement. In so doing, it is important to draw upon evidence focusing on the resilience as well as the vulnerability of populations exposed to political violence, such as Suedfeld has done in reporting on the effects of the Holocaust, Southeast Asian Boat People, and other societal traumas (1997). Furthermore, more research is needed to examine how structural and other variables at the international level are related to traumatic events and their psychological consequences. For example, it is important to develop greater understanding of how individual policymakers' exposure to prior political violence can influence their support for policies in response to subsequent disasters or further instances of political violence. Useful insights may also be gleaned from research on a societal level about the effects of trauma. For example, the relationships between perceived threat, fear, and foreign policy views in the American public after the September 11, 2001, terrorism attacks (Lerner et al. 2003). Hopefully such knowledge can lead to improved political decision making in times of international crisis as well as to a reduction in the long-term impact of traumatic stress on individuals and populations.

As research continues to seek greater understanding of political violence and traumatic stress, the preponderance of evidence compellingly indicates that political violence often leads to severe and enduring psychological harm. Therefore, ethically, experts on the traumatic stress effects of political violence are obligated to convey this knowledge to the public, and political leaders are ethically bound to consider such effects in their decision making. Such responsibilities are all the more challenging because the locus of responsibility for modern disasters is understood to be held by private and public institutions (Boisjoly, Curtis, and Mellican 1989). The ethical challenges of taking action in response to the potential for political violence is complicated further by the fact that even international organizations with humanitarian goals—such as relief agencies—can unintentionally bring about harm in responding to political emergencies and war (Slim 1997). Despite such challenges, however, leaders have an obligation to influence institutions that in turn help to prevent or mitigate the effects of political violence. Leaders also have an obligation to try to build public opinion in support of what they view to be the most ethical course of action in the face of extreme political violence. Such lessons about ethical responsibility of the larger international community have been drawn from examining the Rwandan genocide of 1994 and have potential relevance for responding to genocide in Darfur (Beardsley 2008) as well as to other forms of political violence. The ethical responsibilities entailed in responding to political violence are all the more compelling when considering the traumatic impact of such violence on individuals, groups, and society.

NOTE

A previous version of this paper was presented at the Politics, Psychology, and Ethics Public Forum with members of the International Society of Political Psychology, sponsored by the Interdisciplinary Center for the Scientific Study of Ethics and Morality, in co-sponsorship with the School of Social Science, the Research on International and Global Studies Center, the Department of Political Science, the Center for Global Peace and Conflict Studies, and the University of California, Irvine.

REFERENCES

Afshari, Reza. 2002. "Tortured Confessions: Prisons and Public Recantations in Modern Iran by Ervand Abrahamian (1999)." Review. *Human Rights Quarterly* 24(1): 290–97.

American Psychiatric Association (APA). 2000. *Diagnostic and Statistical Manual of Mental Disorders, Fourth Edition, Text Revision.* Washington, DC: American Psychiatric Association.

Auerhahn, Nanette C., and Dori Laub. 1998. "Intergenerational Memory of the Holocaust." In *International Handbook of Multigenerational Legacies of Trauma,* edited by Yael Danieli, 21–42. New York: Plenum Press.

Baker, Ron. 1992. "Psychological Consequences for Tortured Refugees Seeking Asylum and Refugee Status in Europe." In *Torture and its Consequences,* edited by Metin Basoglu, 83–106. Cambridge, MA: Cambridge University Press.

Beardsley, Brent. 2008. "Learning from the Rwandan Genocide of 1994 to Stop the Genocide in Darfur, Part Two." Accessed October 16, 2008. http://www.journal.dnd.ca/vo6/no2/human-humain-eng.asp.

Beckham, Jean C., Michelle E. Feldman, and Angela C. Kirby. 1998. "Atrocities Exposure in Vietnam Combat Veterans with Chronic Posttraumatic Stress Disorder: Relationship to Combat Exposure, Symptom Severity, Guilt, and Interpersonal Violence." *Journal of Traumatic Stress* 11: 777–85.

Bernsten, Dorthe, Morten Willert, and David C. Rubin. 2003. "Splintered Memories or Vivid Landmarks? Qualities and Organization of Traumatic Memories With and Without PTSD." *Applied Cognitive Psychology* 17: 675–93.

Boisjoly, Russell P., Ellen F. Curtis, and Eugene Mellican. 1989. "Roger Boisjoly and the Challenger Disaster: The Ethical Dimensions." *Journal of Business Ethics* 8: 217–30.

Brewin, Chris R. 2007. Remembering and Forgetting. In *Handbook of PTSD: Science and Practice,* edited by Matthew J. Friedman, Terence M. Keane, and Patricia A. Resick, 116–34. New York: The Guilford Press.

Butler, Lisa D., Cheryl Koopman, Jay Azarow, Christine M. Blasey, Juliette C. Magdalene, Sue DiMiceli, David A. Seagraves, T. Andrew Hastings, Xin-Hua Chen, Robert W. Garlan, Helena C. Kraemer, and David Spiegel. 2009. "Psychosocial Predictors of Resilience after the September 11, 2001 Terrorist Attacks." *Journal of Nervous and Mental Disease* 197(4): 266–73.

Butler, Lisa D., and David Spiegel. 1997. "Trauma and Memory." *Review of Psychiatry* 16: 13–54.

Charney, Dennis S., Ariel Y. Deutch, S. M. Southwick, and J. H. Krystal. 1995. "Neural Circuits and Mechanisms of Post-Traumatic Stress Disorder." In *Neurobiological and Clinical Consequences of Stress: From Normal Adaptation to PTSD,* edited by Matthew J. Friedman, Dennis S. Charney, and Ariel Y. Deutsch, 271–87. Philadelphia, PA: Lippincott-Raven Publishers.

Danieli, Yael E. 1998. "History and Conceptual Foundations." In *International Handbook of Multigenerational Legacies of Trauma,* edited by Yael E. Danieli, 1–20. New York: Plenum Press.

de Jong, Joop T. V. M., Ivan H. Komproe, Mark Van Ommeren, Mustafa El Masri, Mesfin Araya, Noureddine Khaled, Willem van de Put, and Daya Somasundaram. 2001. "Lifetime Events and Posttraumatic Stress Disorder in 4 Postconflict Settings." *Journal of the American Medical Association* 286: 555–62.

de Waal, Alex. 2007. "Dueling over Darfur." *Newsweek.* Accessed October 16, 2008. http://www.newsweek.com/id/69004.

Dimoulas, Eleni, Lisa Steffian, George Steffian, Anthony P. Doran, Ann M. Rasmusson, and C. A. Morgan, III. 2006. "Dissociation during Intense Military Stress is Related to Subsequent Somatic Symptoms in Women." *Psychiatry* 4(2): 66–73.

Dyregrov, Atle, Leila Gupta, Rolf Gjestad, and Eugenie Mukanoheli. 2000. "Trauma Exposure and Psychological Reactions to Genocide among Rwandan Children." *Journal of Traumatic Stress* 13(1): 3–21.

Freedy, John R., Darlene L. Shaw, Mark P. Jarrell, and Cheryl R. Masters. 1992. "Towards an Understanding of the Psychological Impact of Natural Disasters: An Application of the Conservation of Resources Stress Model." *Journal of Traumatic Stress* 5(3): 441–54.

Fremont, Wanda P. 2004. "Childhood Reactions to Terrorism-Induced Trauma: A Review of the Past 10 Years." *Journal of the American Academy of Child & Adolescent Psychiatry* 43(4): 381–92.

Galea, Sandro, Jennifer Ahern, Heidi Resnick, Dean Kilpatrick, Michael Bucuvalas, Joel Gold, and David Vlahov. 2002. "Psychological Sequelae of the September 11 Terrorist Attacks in New York City." *The New England Journal of Medicine* 346: 982–87.

Gerrity, Ellen T., Terrence Martin Keane, and Ferris Tuma. 2001. *The Mental Health Consequences of Torture.* New York: Plenum Publishers.

Ginzburg, Karni, Lisa D. Butler, Kasey Saltzman, and Cheryl Koopman. 2009. "Dissociative Reactions in PTSD." In *Dissociation and the Dissociative Disorders: DSM-V and Beyond* (pp. 457–69), edited by Paul F. Dell and John A. O'Neil. New York: Taylor and Francis.

Ginzburg, Karni, Cheryl Koopman, Lisa D. Butler, Oxana Palesh, Helena C. Kraemer, Catherine C. Classen, and David Spiegel. 2006. "Evidence for a Dissociative Subtype of Posttraumatic Stress Disorder among Help-Seeking Childhood Sexual Abuse Survivors." *Journal of Trauma and Dissociation* 7(2): 7–27.

Greenwald, Glenn. 2007. *A Tragic Legacy: How a Good vs. Evil Mentality Destroyed the Bush Presidency.* New York: Crown Publishers.

Hellawell, Steph J., and Chris R. Brewin. 2002. "A Comparison of Flashbacks and Ordinary Autobiographical Memories of Trauma: Cognitive Resources and Behavioural Observations." *Behavior Research and Therapy* 40(10): 1143–56.

Hamburg, David A., Alexander George, and Karen Ballentine. 1999. "Preventing Deadly Conflict: The Critical Role of Leadership." *Archives of General Psychiatry* 56: 971–6.

Herman, Judith. 1992. *Trauma and Recovery.* New York: Basic Books.

Hjern, Anderes, Birgitta Angel, and Olle Jeppson. 1998. "Political Violence, Family Stress and Mental Health of Refugee Children in Exile." *Scandinavian Journal of Social Medicine* 26(1): 18–25.

Hobfoll, Stevan E. 2001. "The Influence of Culture, Community, and the Nested-Self in the Stress Process: Advancing Conservation of Resources Theory." *Applied Psychology: An International Review* 50(3): 337–421.

Hobfoll, Stevan E., Carla A. Dunahoo, and Jeannine Monnier. 1995. "Conservation of Resources and Traumatic Stress." In *Traumatic Stress: From Theory to Practice,* edited by John R. Freedy and Stevan E. Hobfoll, 29–47. New York: Plenum Press.

Horowitz, Mardi J. 1978. *Stress Response Syndromes.* New York: Jason Aronson.

Hynes, Michelle, and Barbara L. Cardozo. 2000. "Observations from the CDC: Sexual Violence against Refugee Women." *Journal of Women's Health & Gender-Based Medicine* 9(8), 819–23.

Janoff-Bulman, Ronnie. 1992. *Shattered Assumption: Toward a New Psychology of Trauma.* New York: Free Press.

Kamau, Michael, Derrick Silove, Zachary Steel, Ronald Catanzaro, Catherine Bateman, and Solvig Ekblad. 2004. "Psychiatric Disorders in an African Refugee Camp." *Intervention* 2(2): 84–89.

Karunakara, Unni K., Frank Neuner, Margarete Schauer, Kavita Singh, Kenneth Hill, Thomas Elbert, and Gilbert Burnham. 2004. "Traumatic Events and Symptoms of Post-Traumatic Stress Disorder among Sudanese Nationals, Refugees, and Ugandans in the West Nile." *African Health Sciences* 4(2): 83–93.

Kessler, Ronald C., Amanda Sonnega, Evelyn Bromet, Michael Hughes, and Christopher B. Nelson. 1995. "Posttraumatic Stress Disorder in the National Comorbidity Survey." *Archives of General Psychiatry* 52(12): 1048–60.

King, Lynda A., Daniel W. King, John A. Fairbank, Terence M. Keane, and Gary A. Adams. 1998. "Resilience-Recovery Factors in Post-Traumatic Stress Disorder among Female and Male Vietnam Veterans: Hardiness, Postwar Social Support, and Additional Stressful Life Events. *Journal of Personality and Social Psychology* 74(2): 420–34.

Klein, Naomi. 2007. *Shock Doctrine: The Rise of Disaster Capitalism.* New York: Metropolitan Books, Henry Holt and Co.

Koopman, Cheryl, Catherine Classen, and David Spiegel. 1994. "Predictors of Posttraumatic Stress Symptoms among Survivors of the Oakland/Berkeley, Calif. Firestorm." *American Journal of Psychiatry* 151: 888–94.

———. 1996. "Dissociative Responses in the Immediate Aftermath of the Oakland/Berkeley Firestorm." *Journal of Traumatic Stress* 9(3): 521–40.

Kozaric-Kovacic, Dragica, Ana Marusic, and Tajana Ljubin. 1999. "Combat-Experienced Soldiers and Tortured Prisoners of War Differ in the Clinical Presentation of Post-Traumatic Stress Disorder." *Nordic Journal of Psychiatry* 53(1): 11–15.

Kuehn, L. L. 1974. "Looking Down a Gun Barrel: Person Perception and Violent Crime." *Perceptual and Motor Skills* 39: 1159–64.

Kulka, Richard A., William E. Schlenger, John A. Fairbank, Richard L. Hough, B. Kathleen Jordan, Charles R. Marmar, and Daniel S. Weiss. 1990. *Trauma and the Vietnam War Generation: Report of Findings from the Vietnam Veterans Readjustment Study.* New York: Brunner/Mazel.

Latané, Bibb, and John M. Darley. 1968. "Group Inhibition of Bystander Intervention in Emergencies." *Journal of Personality and Social Psychology* 10(3): 215–21.

Latané, Bibb, and Steve Nida. 1981. "Ten Years of Research on Group Size and Helping." *Psychological Bulletin* 89(2): 308–24.

Lerner, Jennifer, Roxana Gonzalez, Deborah Small, and Baruch Fischhoff. 2003. "Effects of Fear and Anger on Perceived Risks of Terrorism: A National Field Experiment." *Psychological Science* 14:144-50.

Loutan, Louis, Paola Bollini, Sandro Pampallona, Donatella Bierens De Haan, and Francoise Gariazzo. 1999. "Impact of Trauma and Torture on Asylum-Seekers." *European Journal of Public Health* 9(2): 93–96.

McCann, I. Lisa, and Laurie A. Pearlman. 1990. *Psychological Trauma and the Adult Survivor: Theory, Therapy, and Transformation.* New York: Brunner/Mazel.

Miller, Arthur G. 2004. "What Can the Milgram Obedience Experiments Tell Us about the Holocaust?" In *The Social Psychology of Good and Evil,* edited by Arthur G. Miller, 193–239. New York: The Guilford Press.

Mills, Edward J., Sonal Singh, Timothy H. Holtz, Robert M. Chase, Sonam Dolma, Joanna Santa-Barbara, and James J. Orbinski. 2005. "Prevalence of Mental Disorders and Torture among Tibetan Refugees: A Systematic Review." *BioMed Central International Health and Human Rights* 5(1): 7.

Modvig, J., J. Pagaduan-Lopez, J. Rodenburg, C. Salud, R. Cabigon, and C. Panelo. 2000. "Torture and Trauma in Post-Conflict East Timor." *The Lancet* 356: 1763.

Mollica, Richard F., Keith McInnes, Thang Pham, Mary C. Smith-Fawzi, Elizabeth Murphy, and Lien Lin. 1998. "The Dose-Effect Relationships between Torture and Psychiatric Symptoms in Vietnamese Ex-Political Detainees and a Comparison Group." *Journal of Nervous & Mental Disease* 186(9): 543–53.

Mollica, Richard F., Keith McInnes, Narcisa Sarajlić, James Lavelle, Iris Sarajlić, and Michael P. Massagli. 1999. "Disability Associated with Psychiatric Comorbidity and Health

Status in Bosnian Refugees Living in Croatia." *Journal of the American Medical Association* 282(5): 433–39.

Munczek, Debora S., and Steven Tuber. 1998. "Political Repression and its Psychological Effects on Honduran Children." *Social Science & Medicine* 47(11): 1699–1713.

North, Carol S., Sara J. Nixon, Sheryll Shariat, Sue Mallonee, J. Curtis McMillen, Edward L. Spitznagel, and Elizabeth M. Smith. 1999. "Psychiatric Disorders among Survivors of the Oklahoma City Bombing." *Journal of the American Medical Association* 282(8): 755–62.

Orr, Scott P., and Roger K. Pitman. 1999. "Neurocognitive Risk Factors for PTSD." In *Risk Factors for Posttraumatic Stress Disorder,* edited by Rachel Yehuda, 125–41. Washington, DC: American Psychiatric Press, Inc.

Pham, Phuong N., Harvey M. Weinstein, and Timothy Longman. 2004. "Trauma and PTSD Symptoms in Rwanda: Implications for Attitudes toward Justice and Reconciliation." *The Journal of the American Medical Association* 292(5): 602–12.

Pratchett, Laura, Lisa M. Brown, and Bruce Bongar. 2006. "Reflections on the Psychology of Terrorism." In *Psychology of Terrorism,* edited by Bruce Bongar, Lisa M. Brown, Larry E. Beutler, James N. Breckenridge, and Philip G. Zimbardo, 452–57. New York: Oxford University Press.

Resnick, Heidi S., Rachel Yehuda, Roger K. Pitman, and David W. Foy. 1995. "Effect of Previous Trauma on Acute Plasma Cortisol Level Following Rape." *American Journal of Psychiatry* 152: 1675–77.

Rice, Condoleezza. 2002. "Acknowledge That You Have an Obligation to Search for the Truth." *Stanford Report,* June 16. Accessed November 27, 2008. http://news-service.stanford.edu/news/2002/june19/comm_ricetext-619.html.

Rothe, Eugenio M., John Lewis, Hector Castillo-Matos, Orestes Martinez, Ruben Busquets, and Igna Martinez. 2002. "Posttraumatic Stress Disorder among Cuban Children and Adolescents after Release from a Refugee Camp." *Psychiatric Services* 53: 970–89.

Schaal, Susanne, and Thomas Elbert. 2006. "Ten Years after the Genocide: Trauma Confrontation and Posttraumatic Stress in Rwandan Adolescents." *Journal of Traumatic Stress* 19(1): 95–105.

Schuster, Mark A., Bradley D. Stein, Lisa H. Jaycox, Rebecca L. Collins, Grant N. Marshall, Marc N. Elliott, Annie J. Zhou, David E. Kanouse, Janina L. Morrison, and Sandra H. Berry. 2001. "A National Survey of Stress Reactions after the September 11, 2001, Terrorist Attacks." *The New England Journal of Medicine* 342: 1507–12.

Shalev, Arieh Y. 1999. "Psychophysiological Expression of Risk Factors for PTSD." In *Risk Factors for Posttraumatic Stress Disorder,* edited by Rachel Yehuda, 143–61. Washington, DC: American Psychiatric Press, Inc.

Silove, Derrick, Zachary Steel, Patrick McGorry, Vanessa Miles, and Juliette Drobny. 2002. "The Impact of Torture on Post-Traumatic Stress Symptoms in War-Affected Tamil Refugees and Immigrants." *Comprehensive Psychiatry* 43(1): 49–55.

Slim, Hugo. 1997. "Doing the Right Thing: Relief Agencies, Moral Dilemmas and Moral Responsibility in Political Emergencies and War." *Disasters* 21(3): 244–57.

Solomon, Zahava. 1989. "A 3-Year Prospective Study of Post-Traumatic Stress Disorder in Israeli Combat Veterans." *Journal of Traumatic Stress* 2(1): 59–73.

Somasundaram, Daya. 2002. "Child Soldiers: Understanding the Context." *British Medical Journal* 324: 1268–71.

Sourander, Andre. 1998. "Behavior Problems and Traumatic Events of Unaccompanied Refugee Minors." *Child Abuse and Neglect* 22(7): 719–27.

Speckhard, Anne, and Khapta Akmedova. 2004. "Mechanisms of Generating Suicide Terrorism: Trauma and Bereavement as Psychological Vulnerabilities in Human Security—The Chechen Case." In *NATO Science Series,* edited by Jill E. Donnelly, 59–64. Fairfax, VA: IOS Press Inc.

Spiegel, David, and Etzel Cardeña. 1991. "Disintegrated Experience: The Dissociative Disorders Revisited." *Journal of Abnormal Psychology* 100(3): 366–78.

Staub, Ervin. 1999. "The Origins and Prevention of Genocide, Mass Killing, and Other Collective Violence." *Peace and Conflict: Journal of Peace Psychology* 5(4): 303–37.

Suedfeld, Peter. 1997. "Reactions to Societal Trauma: Distress and/or Eustress." *Political Psychology* 18(4): 849–61.

Tromp, Shannon, Mary P. Koss, Aurelio J. Figueredo, and Melinda Tharan. 1995. "Are Rape Memories Different? A Comparison of Rape, Other Unpleasant, and Pleasant Memories among Employed Women." *Journal of Traumatic Stress* 8(4): 607–27.

Tucker, Phebe, Betty Pfefferbaum, Sara J. Nixon, and Warren Dickson. 2000. "Predictors of Post-Traumatic Stress Symptoms in Oklahoma City: Exposure, Social Support, Peri-Traumatic Responses." *The Journal of Behavioral Health Services & Research* 27(4): 406–16.

United Nations Secretary General. 2008. "With 16 Million Refugees Worldwide, Secretary-General Calls for Redoubled Efforts to Address Causes, Consequences; More Equitable Sharing of Burden of Protection." Accessed July 1, 2008. http://www.un.org/News/Press/docs/2008/sgsm11643.doc.htm.

Weine, Stevan M., Daniel F. Becker, Dolores Vojvoda, Emir Hodzic, Marie Sawyer, Leslie Hyman, Dori Laub, and Thomas H. McGlashan. 1998. "Individual Change after Genocide in Bosnian Survivors of 'Ethnic Cleansing': Assessing Personality Dysfunction." *Journal of Traumatic Stress* 11(1): 147–53.

Yehuda, Rachel, Boaz Kahana, Karen Binder-Brynes, Steven M. Southwick, John W. Mason, and Earl L. Giller. 1995. "Low Urinary Cortisol Excretion in Holocaust Survivors with Posttraumatic Stress Disorder." *American Journal of Psychiatry* 152(7): 982–86.

Zimbardo, Philip. 2007. *The Lucifer Effect: Understanding How Good People Turn Evil.* New York: Random House.

Zoroya, Gregg. 2008. "A Fifth of Soldiers at PTSD Risk: Rate Rises with Tours, Army Says." *San Francisco Chronicle,* March 7.

Chapter 8

About Law and Justice

Gilbert Geis and Bridgette Portman

Conversation

Bridgette Portman (BP): To start off, could you tell me a bit about yourself and your background, and how you became interested in criminology?

Gil Geis (GG): I was born in Brooklyn, New York, in 1925, which makes me eighty-four—that surprises me every morning. I was in the first class of an extraordinarily good high school, the Bronx High School of Science. In fact, the kid who sat next to me won the Nobel Prize in physics a couple of years ago. It was an unbelievable school. Then I went into the Navy during the Second World War, and one of the things that made me an unrepentant patriot, even though I often hate what the government is doing, is that after I spent a year and a half in Brazil, they sent me to college. And after the war, I got the GI bill, which paid for five more years of college, and I am unbelievably grateful to the government for that. They sent me to Colgate University, which never would have admitted me—I had a lousy high school record. And then after the war I went to Stockholm, Sweden, came back and went to the University of Wisconsin and got my PhD there in sociology. It was very, very difficult to get work at that time, even with a pretty decent degree, and I think I got about three job possibilities in the whole country. I took the University of Oklahoma and went there for five miserable years. They simply asked me to teach criminology and that's how I got in; I never had a criminology course in my whole life.

I grew up quite left wing; my high school was left wing—the faculty was probably half communist—and I got particularly interested in what you would call white-collar crime. Whatever reputation I have is in that particular field. I was a newspaper reporter for quite a number of years, and I write very easily. I've written four hundred and something articles, which is quite a lot of articles, if anybody tried to read them. And this case [the Abu Omar case, discussed in the following investigation] seemed a natural one for me. Joe DiMento and I had collaborated on probably seven or eight different articles. He speaks Italian, he was in Milan, and he became personally acquainted with the prosecutor in the case. When it first broke, I thought it was really an extraordinary kind of event—extraordinary rendition *was* extraordinary—and Joe had the original sources from there. We collaborate very amiably. Ultimately, we should have done a book, but we each got caught up in other kinds of endeavors. We're still interested in the subject. It's absolutely boomeranged. I noticed [CIA Director] Leon Panetta, for example, was asked whether he'd continue the practice of rendition. I would have presumed he'd say no, never, we're not going to do that anymore. But he said he wasn't categorically going to stop it. So that's where we get into what Kristen [Monroe] is interested in, the moral issue—whether extraordinary rendition is worth the moral abhorrence that it brings upon the nation in terms of what you're going to get for it, if anything.

BP: It brings up a lot of complicated ethical issues. Somebody could argue, for example, that it's justified if it results in making America safer. How would you respond to that?

GG: You put your finger on exactly the moral issue. I spent a year at Harvard Law School, and the professor would say, "Now who believes in torture?" Everyone in the class was absolutely indignant; they didn't believe in torture. And then of course he would pose the hypothetical: If in fact we know that somebody has planted an atomic bomb in the subway station somewhere in Manhattan, and it's going to explode in X amount of time, should we torture the man that we captured who knows where the bomb is? And then the law students become very confused, and by the time they're seniors they never answer anything categorically. That's what makes them good lawyers; they never take a position. But that's the issue. How much of the integrity and dignity and decency of the nation are you willing to sacrifice in a program such as this, in order to hypothetically protect the citizens?

As another illustration, John Yoo, the man who wrote the Department of Justice memorandum that defended torture—he's teaching at Chapman now on a visiting deal—says we've staved off a Mumbai replication by using waterboarding. Well, I don't know if that's true, and I don't know if it's false. I wrote a book on witchcraft—a pretty good book, all things being equal—in England in the seventeenth century.[1] And one thing that I agree with John McCain about, though I don't totally agree with him, is that torture gets you a lot of phony information because people ultimately come to the point where they'll say anything in order to be relieved of the agony of it. But on the other hand, that's not totally correct. Sometimes it will get you accurate information. Now in the witchcraft cases, of every woman who was tortured and asked to admit that she was a witch, none of them were witches, not a single one. And they burned and hanged thousands of people, on confessions.

BP: And they all admitted it.

GG: Yeah. "Did you kill his cattle?"—"Oh yeah, I killed his cattle." "Did you put a curse on her that kept her from having children?"—"Yeah, that was me." They admitted to things that today we appreciate are totally impossible. And they were tortured. They weren't tortured in England, they were tortured on the continent. England did not torture, for some peculiar reason. The only way you could torture was with the king's permission, and the king was not willing to give that permission. There's a very good book by John Langbein on the history of torture.[2] It's very interesting.

BP: Extraordinary rendition was practiced on a large scale under the Bush administration. Now we have the Obama administration. Do you think that it's going to be a practice that is stopped?

GG: Well, that's essentially what I was talking about with Panetta. During his confirmation hearing, he said he certainly was not going to use it as much. And he also said that they're going to be very, very careful that the places where they send these people do not torture. But, you know, once you send them there you have no control over it. So I don't know whom Panetta was trying to placate. I would have thought he would say, "No, we're not going to do that. That's something the United States doesn't stand for, something we don't believe in." Or he could take the position that it wouldn't be useful. None of those are necessarily empirically defensible, but he didn't, which really surprised me. I was really quite surprised. Another wonderful illustration—I listened to Richard Cheney saying that he believed in waterboarding and had no regrets about using it. And then he had the nerve to say, well, the Department of Justice attorneys approved it, so obviously it was legal. But all of those people were appointed very carefully by Cheney and screened for the fact that they would approve waterboarding. So you have a biased opinion that virtually everybody who's read those memorandums thinks is very, very superficial.

BP: What do you think it will take to stop this practice?

GG: Gee, that's a nice question. I think what tends to stop virtually anything is when a case arises that is so egregious, so awful, that it focuses people's attention. Extraordinary rendition is going on, but people really don't pay much attention to it. They've made movies out of it, there have been a few books written, but they don't pay a lot of attention to it. It's too general, too vague. It has nothing to do with them. But if you get a case that really comes to the courts, and somebody with power and media contacts highlights it, then you can arouse public sentiment against it. Virtually all of the decent laws—the FDA is a perfect illustration—arise out of some awful situation that awakens people.

BP: What did you think about Obama's decision to close down Guantanamo?

GG: Oh, I think that's delightful. I've been in Guantanamo—not in the prison, I was there during the war, when it was a naval station. I spent an evening and a day there. It strikes me—and I tend to go into hyperbole on those sorts of things—that to put people into Guantanamo and then claim that they're not on American soil, therefore they're entitled to no rights, is the utmost kind of hypocrisy. It doesn't make any sense to me. They're American prisoners, and they're entitled to the rights of any American prisoners. Guantanamo has become a symbol now of some very wrong things in the United States, and I think it ought to be closed. I don't know why the Cubans ever let us keep it. But we got it in a treaty, and they can't break the treaty.

BP: Extraordinary rendition maybe can be seen as part of a larger tendency for the government to put security, or what it sees as security, above civil liberties. We've also had things like indefinite detention of enemy combatants, wiretapping, the Patriot Act.

GG: Yeah, the whole scenario. But the case that Joe and I wrote about is so awful. They pick this guy up, and they go off to Venice, or wherever they went to, and spend thousands of dollars celebrating what they did. It's something that is so offensive it just revolts you. Now again, you raised the question at the beginning, does it serve a superior purpose? I can't answer that question satisfactorily. It seems to me that you can sacrifice a certain degree of safety for principle, essentially. The illustration that comes to mind is when sex offenders are released and they have to notify the neighborhood that you have a sex offender in your midst. Then everybody goes out and holds signs up and harasses them, and my feeling is that that behavior is so uncalled for and so nasty that I'd rather have my kid run a very mild risk of being the victim of a sex offender, however awful that might be, than see this demonstration of public ugliness and antipathy toward a human being. But you know, if it were my kid, on the other hand, I'm not sure I'd have such noble sentiments. That's always the problem.

BP: There are also issues here, of course, of international law. Rendition is a violation of the UN convention against torture.

GG: Oh, it's clearly a violation of international law. But you know, Bush couldn't care less about international law.

BP: So is part of this a larger issue of the US not being willing to honor its international commitments?

GG: Yeah, or not being willing to adhere to international commitments. You know, the imperial presidency. Bush rejects the international court in The Hague, and goes against the Kyoto Treaty, and there was the pre-emptive invasion of Iraq. It's ghastly. You don't do those sorts of things in a civilized, decent, democratic country. But we did them.

BP: Moving on to a different topic, you've done a lot of work on white-collar crime. I have to ask you what you thought of the Bernard Madoff case. Recently he pleaded guilty and he's been sent to jail.

GG: What do I think about the Madoff case? It's just a standard Ponzi scheme. One is perplexed that the SEC never toppled it. They had constant warnings, very sharp and strong warnings, that he was up to something, that what he was doing was not legitimate, that there was a hoax going on, and they never investigated it. That he did it? Greed is something that I don't really understand very well. How much do you need? When Merrill Lynch gives seven hundred employees a million-dollar bonus while they're going down the drain, I don't understand that. When an AIG guy gets a $12 million dollar buyout when he was getting a salary of a million—what does he do with all that money? If I had it, I don't know what I'd do with it. So you put it in a Swiss bank, I guess. Then you don't have to pay taxes on it.

BP: Do you think it's an issue of individual people's greed, or . . .

GG: It's systemic. It's a capitalistic society. As I said very early, I'm enormously indebted to the United States, but it has some very serious flaws. It's a greedy, capitalistic society in which all the pressure is to get money, get more, consume. Success is measured by income. "If you're so smart, how come you're not rich?" is a very common phrase. And as you watch the debate going on now over Obama's economic moves, "socialization" becomes a terribly nasty word. They're going to "socialize" things. The whole breakdown of Wall Street is, again, greed. You've got a million dollars but the guy next door has a million and a half, you've got a Toyota and he's got a Jaguar, and you've got to compete with him. It's very intense. The pressure is enormous to get rich, and ultimately, there are only a few ways to get rich. One is to be legal, and the easier way is to be a crook. And it's very tempting.

BP: So what's the solution? We need more regulation?

GG: Well you certainly need more supervision, you need more transparency, but that's almost a cliché. The solution is that there be a much more intense education in the business schools and in the medical schools toward honesty, decency, compassion, caring, and much less selfishness. The United States has always been an individualistic society. The best illustration I can offer is that in American law, if somebody needs help, and I can offer that help without any risk to myself, I'm not required to do so. There is no European country that doesn't mandate that you act as a Good Samaritan. And if you don't act as a Good Samaritan, they're going to put you in prison. If I'm standing there and a blind man comes along and he's likely to walk off a cliff, and I wave bye-bye, and he walks off the cliff, I have no legal responsibility for that.

BP: Because you're not actively causing harm?

GG: I haven't caused it, I'm not related to him, and I have no contract to protect him; those are the three conditions under which I've got to say, "You're going to walk off the cliff and kill yourself." Now in France, they arrest people constantly for not being Good Samaritans. Benazir Bhutto, who just got assassinated—her brother got picked up, I think, as a non-Good Samaritan [it was her brother's wife, Rehana Bhutto]. Now in the United States, doctors are particularly exempt from liability, even if they stop on the road and try to assist somebody. But you're liable if you aren't successful, so people don't intervene. Very individualistic society.

BP: Before the interview, you mentioned you're working on a book right now. What is it about?

GG: The book I'm just finishing is based on the theme that power is an aphrodisiac. I've done four chapters now on case studies of congressmen who have gotten into trouble erotically: Mark Foley from Florida, who was harassing pages in the Senate; Larry Craig, who was picked up on a charge in the Minneapolis-Saint Paul airport for soliciting a policeman; Bob Packwood, who had a strong tendency to come on sexually to women working with him or otherwise encountered; and Eliot Spitzer, who was involved in the prostitution case. And then in the introductory chapter I'm going to deal with Kennedy and Johnson, and Clinton of course.

BP: That's pretty interesting.

GG: You know, I've written about twenty-six books. I finally decided that since I'm not working, and I don't need to be promoted, I'm going to try to write one that is fun. I wrote fiction when I was a kid. I wrote a couple of detective books. That was fun. But you can't get ahead in academia writing detective books.

BP: Do you have a favorite out of all the books and things that you've written, something that was most enjoyable to work on?

GG: I wrote one with a graduate student, Mary Dodge, on the fertility clinic scandal at Irvine.[3] It was a "slightly dangerous" enterprise—they don't love me; the administration was not exactly thrilled. I thought they were extremely culpable of a lot of things. Some doctors were taking eggs from woman X and planting them in woman Y with nobody's consent. I liked that book. Then I wrote one that's much like the sex book, only on a higher level, called *Crimes of the Century,* in which I took up five different crimes—Leopold and Loeb, Scottsboro, O. J. Simpson—I liked that book.[4] The witch book I worked on for endless years, and I'm very fond of that one. I write; it's a habit.

NOTES

1. Geis and Bunn 1997.
2. Langbein 1977.
3. Dodge and Geis 2003.
4. Geis and Bienen 1998.

REFERENCES

Dodge, Mary, and Gilbert Geis. 2003. *Stealing Dreams: A Fertility Clinic Scandal.* Boston, MA: Northeastern University Press.

Geis, Gilbert, and Leigh B. Bienen. 1998. *Crimes of the Century: From Leopold & Loeb to O. J. Simpson.* Boston, MA: Northeastern University Press.

Geis, Gilbert, and Ivan Bunn. 1997. *A Trial of Witches: A Seventeenth-Century Witchcraft Prosecution.* New York: Routledge.

Langbein, John H. 1977. *Torture and the Law of Proof: Europe and England in the Ancien Régime.* University of Chicago Press.

Ethics and the Policy of Extraordinary Rendition

Joseph F. C. DiMento and Gilbert Geis

Investigation

On February 17, 2003, forty-two-year-old Moustafa Hassan Nasr, commonly known as Abu Omar, an imam from Egypt, was allegedly kidnapped by American and Italian officials and flown from Italy through Germany to Egypt, where he was subjected to a brutal regime of torture. When we first discussed the Omar case in 2006, we viewed it as an unusual action by agents of the US government (DiMento and Geis 2006). However, what happened to Abu Omar was the commonplace fate of many people whom US agents or accomplices transported to overseas sites so that their interrogations would not be associated with American policy. Legal judgments of this practice, known as "extraordinary rendition" and sometimes labeled as "torture by proxy," have reached varying conclusions. For our part, we regard extraordinary rendition as immoral and as a violation of both American and international law. In this chapter, we address legal, policy, and moral issues bearing on extraordinary rendition and emphasize that there is no evidence that the practice produces useful intelligence and that even if it were to do so, the means being used to that end are intolerable. We also believe that while some might seek to justify the policy in jurisprudential terms, it is a legally highly suspect and morally reprehensible procedure that brings discredit and shame upon the United States.

ILLUSTRATIVE EPISODES

The Milano Case

Abu Omar, after stays in Albania and in Munich, had gained political asylum in Italy. He faced prosecution if he returned to Egypt because of his onetime membership in a radical Islamic organization, Jamaat al Islam. Omar also had been a government informant for SHIK, the Albanian intelligence agency, and there was speculation that the Italians and/or the Americans might desire to induce him to play the same role in Italy in regard to possible Muslim terrorists. At the time Omar was kidnapped the anti-terrorist branch of the Italian police (DIGOS) was investigating him; five of his colleagues were already on trial before an Italian court.

Walking to noon prayer at a mosque on Via Guerzoni in Milan, Omar was confronted by men dressed as Italian police officers. One of those men later told prosecutors that Omar was lured to a van by asking him to show his identification papers. The man providing these details testified that he had been assured by the Central Intelligence Agency (CIA) station chief in Milan, Robert Seldon Lady, that the operation had "cover."[1] Numerous CIA agents were involved in the planning and execution of the rendition, if one accepts the facts presented in the Italian case brought against American and Italian operatives.

The kidnappers sprayed chemicals into Omar's face and thrust him into the minivan. He was then taken to Aviano, a joint American-Italian air base northeast of Venice, then to the Ramstein air base in Germany. From there he was flown to Egypt. There he allegedly underwent a severe torture regimen that included electric shocks to his genitals, being hung upside down, being bombarded with loud noises, and being moved from a hot sauna into a refrigerated cell. The credibility of Omar's description of the regimen to which he was subjected is reinforced by a report by the US Department of State (2003) on human rights violations in Egypt:

> Principal methods of torture reportedly employed ... included victims being: stripped and blindfolded; suspended from a ceiling or doorframe with feet just touching the floor; beaten with fists, whips, metal rods, or other objects; subjected to electric shocks; and doused with cold water.... Some victims, including male and female detainees and children, reported that they were sexually assaulted or threatened with rape themselves or family members. (§1c)

Roberto Castello, the justice minister in the Italian government of Silvio Berlusconi, refused to initiate extradition proceedings against the defendant CIA agents, claiming that matters of national security would be compromised if he did so. Castello delayed his decision until April 2006, just after his party had lost the general election. The victorious center-left government of Romano Prodi also denied an extradition request. A challenge to the trial's legality led to its suspension in June 2007, ten days after it had gotten underway. The cessation was implemented to allow the Italian constitutional court to rule on whether the prosecutors had acted beyond their authority by employing documents classified as state secrets. The decision by the constitutional court in March 2009 was a partial victory for both sides. The prosecutor has written that the trial will proceed with certain evidence suppressed but with sufficient evidence to establish culpability.

El-Masri v. United States

Khaled El-Masri, Lebanon-born but a German citizen by marriage, was on vacation in the Balkans when he was removed from a bus at a border crossing into Macedonia on New Year's Eve in 2003. It would later turn out that he had been mistaken for a member of the Hamburg al-Qaeda cell with the name Khalid al-Marsi. El-Masri alleged that after detention by the Macedonian police for three weeks, he was blindfolded and driven to what sounded like an airstrip that lay approximately one hour from Skopje. In a court filing, he claimed that he saw seven or eight men there who were dressed in black and wore black ski masks. He believed that these were members of what is known as the CIA "black renditions" team, a group that in El-Masri's judgment was operating illegally at the direction of George Tenet, the then-CIA director. His captors put a diaper on El-Masri, dressed him in a tracksuit and earmuffs, and dragged him onto an airplane. He later discovered that he had been flown to Kabul; he was held in solitary confinement there for almost five months, where he received meager food rations and putrid water. On his release, he was dropped off in a desolate region in Albania. El-Masri's claims would be examined in a June 2006 Council of Europe report on American rendition practices. The report concluded that his allegation that he was beaten, stripped of his clothing, drugged, and sodomized was substantially accurate.

El-Masri filed suit in a US court alleging violations of his US Constitution Fifth Amendment guarantee of the right to due process. He charged that the United States had contravened the prohibition against subjecting any person to treatment that shocks the conscience and that punished an individual without legal process. The suit also alleged violations of the Alien Tort Statute and of the international legal norm against arbitrary and prolonged detention.

A Court of Appeals held that the state security concerns dictated that El-Masri could not mount a discovery proceeding and that the case had to be dismissed because the government's charges could not be examined in their full depth. The ruling was based on the state secrets doctrine that provides a daunting challenge to adjudication of claims of extraordinary rendition. As of 2007, it had been invoked thirty-nine times in the previous six years compared with fifty-nine times in the previous twenty-four years. The privilege holds that the United States may prevent the disclosure of information in a judicial proceeding if "there is a reasonable danger" that such disclosure "will expose military matters which, in the interest of national security, should not be divulged." The El-Masri court relied on the leading Supreme Court case, *United States v. Reynolds,* and on what it said were principles "well established in the law of evidence," tracing the origin of such doctrine to the 1807 treason trial of Aaron Burr.

The ruling in the El-Masri case raises a significant obstacle to challenges of government misconduct, particularly in times of warfare. The decision noted that "[t]he courts are ill-equipped to become sufficiently steeped in foreign intelligence matters to serve effectively in the review of secrecy classifications in that area." The idea that alleged relevant state secrets might be examined in camera or by a designated impartial party or parties was rejected. The judge in the El-Masri case stated: "When ... the occasion is appropriate ... the court should not jeopardize the security which the privilege is meant to protect by insisting upon the examination of the evidence, even by the judge alone, in chambers." Performance evaluations of CIA operatives and the identities of pertinent witnesses were held to be state secrets. El-Masri contended that the information he sought was not truly a state secret, but the court observed that if his civil action were permitted to proceed "the facts central to its resolution would be the roles, if any, played by the defendants in the events he alleges." To establish a prima facie case, the opinion declared, El-Masri would "be obliged to provide admissible evidence not only that he was detained and interrogated, but that the defendants were involved ... in a manner that renders them personally liable to him." The court reasoned that to demonstrate this, El-Masri would necessarily have to expose "how the CIA organizes, staffs, and supervises its most sensitive intelligence operations."

The judge believed that El-Masri and his attorneys misunderstood the relationship of the courts to the executive branch, that the judiciary does not have "a roving writ to ferret out and strike down executive excess," and that its role is more modest. "Article III [of the United States Constitution], however, assigns the courts a more modest role: we simply decide cases and controversies."

Left unaddressed was the question regarding who, if not the judiciary, had the authority to rein in "executive excess." Certainly it is not likely to be accomplished by a Congress that shares party loyalty with the executive. Critics of the outcome of the El-Masri decision noted that the camouflage of the state secrets doctrine was often employed by an administration to avoid embarrassment, to handicap political enemies, and to prevent criminal investigation of executive actions. It was emphasized that when later-declassified information was examined, it had been demonstrated that what were said to have been state secrets were not, most notably in regard to the revelations in the Pentagon papers. For his part, El-Masri said in an interview that his experience with the Americans led him to conclude that "this is not democracy.

In my opinion this is how you establish a dictator regime. Freedom and justice are disrespected, as are basic morals and values" (El-Masri 2005).[2]

The court did recognize that "when the state secrets privilege is validly asserted, the result is unfairness to individual litigants—through the loss of important evidence or dismissal of a case—in order to protect a greater public value." And it recognized the gravity of its conclusion that El-Masri must be denied a judicial forum for his complaint: "dismissal on state secrets grounds is appropriate only in a narrow category of disputes."

In October 2007, the US Supreme Court, in a one-line order, refused to hear El-Masri's appeal, letting stand the appeals court ruling without comment.[3]

Arar v. Ashcroft

Syrian-born Maher Arar, a telecommunications engineer, had lived in Canada for seventeen years. In September 2002, during a layover in John F. Kennedy airport in New York en route to Montreal from Tunis, he was seized by Federal Bureau of Investigation (FBI) and Immigration and Naturalization (INS) agents and held in a detention center for twenty days. After that, US officials ordered that he be transported to Syria in an act of extraordinary rendition. They contended that his removal there was consistent with Article 3 of the United Nations Convention against Torture and Other Cruel, Inhuman, or Degrading Treatment or Punishment (CAT). Arar alleged that in Syria he was placed in a "grave," a cell six-feet long, seven-feet high, and three-feet wide. The cell was damp and cold, contained very feeble light, and was infested with rats. Cats would urinate on Arar through an aperture in the ceiling, and sanitary facilities were non-existent. The prisoner was allowed to bathe in cold water once a week. He was prohibited from exercise and he claimed that the food he was served was barely edible. Arar said that he was beaten on various parts of his body with a two-inch-thick shredded cable and threatened with placement in a spine-breaking chair and with electric shocks. He lost forty pounds during his ten-month period of captivity in Syria.

To alleviate the torture, Arar said he falsely confessed to having trained with terrorists in Afghanistan, even though he never had been in the country and had never been involved in terrorist activity. "I was willing to do anything to stop the torture," Arar said later. He was released on October 10, 2003, with no charges having been filed against him in any jurisdiction.

Arar filed a lawsuit in the United States. The appellate court for the second circuit dismissed the case on the ground that as a non-citizen, Arar lacked standing to bring a claim under the Torture Victim Protection Act against government agents in their official capacities. Furthermore, Arar was said to have no cause of action under the Constitution's due process clause in light of the national security and foreign policy considerations at stake. Again, state security was invoked to keep Arar from litigating his claims. A rehearing was held before all twelve members of the court in December 2008. A Commission of Inquiry in Canada supported Arar's allegations and the government awarded him $10 million Canadian dollars as compensation for his suffering.[4]

Mohammed et al. v. Jeppesen Dataplan, Inc.

In 2007, the American Civil Liberties Union (ACLU), on behalf of five former detainees who alleged that they were subject to extraordinary rendition, brought suit against Jeppesen Dataplan, Inc., a Boeing subsidiary with headquarters in Colorado that was involved in the planning of transport to foreign countries as part of the extraordinary rendition program. Allegedly, Jeppesen had intentionally submitted dummy flights to aviation authorities to camouflage their actual flying itinerary. The company's managing director was quoted as saying: "We do all of the extraordinary rendition flight—you know, the torture flights. It certainly pays well" (Mayer 2008, 129). The state secrets privilege was claimed by the government, which had intervened in the case. At the district-court level, the privilege formed the basis for denying the plaintiffs' action. On appeal, however, the ninth circuit court of appeals concluded that the subject matter of the lawsuit was not a state secret and allowed the case to go forward.

We could provide further examples of extraordinary rendition, but the foregoing should suffice to give a sense of what has been occurring on a considerable scale. In February 2007, for instance, the European Parliament adopted a report naming fourteen nations that had been involved in more than twelve hundred CIA flights. These included Britain, Germany, Ireland, Portugal, Poland, and Romania, with the latter two also allowing the presence of CIA secret detention facilities on their territory.

The United States has adopted the official and often-repeated position, despite evidence to the contrary, that its policies preclude extraordinary rendition. The record indicates that the practice began to be used on a limited basis during the Reagan administration and apparently during the Clinton years as well. It supposedly in those times was "subject to strict procedures" including the need for approval by the other governments involved (Weissbrodt and Bergquist 2006, 123). After September 11, 2001, the pace of extraordinary rendition increased dramatically. From that day through the middle of 2005, there has been an estimated total of sixty to seventy cases involving Egypt alone, and perhaps another sixty involving Jordan, Yemen, Morocco, Pakistan, and Uzbekistan.[5]

THE LAW OF EXTRAORDINARY RENDITION

The four brief profiles of instances of extraordinary rendition outlined above indicate the legal hurdles that confront victims of the practice. Perplexing issues abound.

One wonders what laws plaintiffs might rely on if the preliminary jurisprudential obstacles could be overcome. And would the courts find the practice legal?

Answers to these questions are neither self-evident nor unequivocal. Part of the uncertainly lies in the widespread absence of respect for international law, including among American attorneys and legal scholars. Another part is inherent in the nature of the American legal system, a matter to be addressed more fully later in this chapter. And to a somewhat lesser, but still significant extent, a third element of the problem arises from the intensity and depth of the reaction in the United States to the September 11 catastrophe. What might have proven to be a more clear-headed and, perhaps, a more consensual legal analysis before the 2001 attacks became a highly politicized and emotional set of conditions. We are told that we need to "work through, sort of, the dark side," that "you are either with us or against us," and that "now the gloves come off." These are not comments by bullies on playgrounds or by street thugs; they are uttered by elected and appointed officials running the US government.

We need initially to determine precisely what extraordinary rendition is, a task critical to a consideration of its legality. There are two common ways of defining the practice, the first a generic description, the second based on American practice. Our analysis will be based on both of these characterizations of extraordinary rendition.

1. A suspect is captured or kidnapped abroad by agents of one nation and rendered over to another foreign entity for interrogations that are not constrained by restrictions found in the initiating state.
2. Agents of the United States abduct or arrange for the abduction or forcibly apprehend persons in foreign countries who they suspect of terrorist activities. These persons are then taken to other nations where they are tortured by local enforcement personnel in order to secure information that the Americans desire.

A core element of the definitions is that the practice involves torture that, as we shall see, is itself subject to varying and controversial definitions. Legal scholars usually conclude that kidnapping would pose no great legal challenges under American law, but the use of torture would raise serious legal questions. But because extraordinary rendition also involves kidnappings that occur on the soil of a sovereign nation with or without the express authority from that nation, however that might be established, a further legal—indeed, criminal—element comes into play.

It is legally relevant whether the subject is rendered to a foreign site or held in a place within American jurisdiction. When the case involves incarceration in the prison at Guantánamo Bay in Cuba, a strip of land about the size of Manhattan, the legally complex question becomes whether the suspect is on American soil or on soil outside the reach of domestic or international law binding on the United States. Reportedly the practice of rendition also involves forcible return of a Guantánamo detainee to countries where they are at risk of torture (United Nations Commission on Human Rights 2006).[6]

The definition of torture also is crucial for legal analysis of extraordinary rendition. One viewpoint is that brutal techniques beyond what is permissible under the Army Field Manual (AFM) guidelines constitute torture. Others maintain that neither the AFM nor international laws and treaties establish the limits of what are acceptable tactics. But when it comes to specific forms of torture, it often is arguable how they reasonably ought to be regarded. For instance, whether waterboarding, a simulated drowning procedure, is torture has voluble advocates on both sides of the question.[7]

Most directly bearing on the matter of general issues is the United Nations CAT, adopted by the General Assembly in 1984 and entered into force in June 1987. As of December 2008, 146 countries, including the United States, had ratified the convention. It obliges signatories to prohibit and to punish torture as well as cruel, inhuman, and degrading treatment in all circumstances. The convention compels ratifying governments to investigate all allegations of such behavior and to provide remedies for their victims.

Article 3 of the convention is particularly relevant to the practice of extraordinary rendition. It prohibits ratifying states from expelling, returning, or extraditing a person "to another State where there are substantial grounds for believing that he would be in danger of being subject to torture."

Torture is described under the treaty in the following terms:

> ... any act by which severe pain or suffering, whether physical or mental, is intentionally inflicted on a person for such purposes as obtaining from him or a third person information or a confession, punishing him for an act he or a third person has committed or is suspected of having committed, or intimidating or coercing him or a third person, or for any reason based on discrimination of any kind, ... when such pain or suffering is inflicted by or at the instigation of or with the consent or acquiescence of a public official or other person acting in an official capacity. "Torture" does not include pain or suffering arising only from, inherent in or incidental to lawful sanctions.

Article 2(a) of the convention adds: "No exceptional circumstances whatsoever, whether a state of war or a threat of war, internal political instability or any other public emergency may be invoked as a justification of torture, or other cruel, inhuman or degrading treatment or punishment."

The Geneva Conventions also are relevant. The conventions include Geneva III, the Convention Relative to the Treatment of Prisoners of War, and Geneva IV, the Convention Relative to the Protection of Civilian Persons in Time of War. They address the treatment of individuals caught up in military activities in a wide variety of ways.

> International humanitarian law distinguishes "lawful" combatants from other combatants and civilians. Those participating in hostilities who are not lawful combatants are "belligerent persons who lack the privilege enjoyed by the armed forces of a state to engage in warfare with immunity from any liability under national or international law, except as prescribed by the international law of war. . . ." The term "unlawful combatant" is commonly used to refer to a person "taking direct part" in an international armed conflict "without being entitled to do so." A "lawful" combatant is entitled to prisoner of war (POW) status when captured, and hence to the full protection of the (Third) 1949 Geneva Convention Relative to the Treatment of Prisoners of War.... These protections include ... the right not to be interrogated without consent (and the associated right not to be subject to torture or coercion to secure information ...) POWs include ... as well ... members of militias or volunteer corps forming part of such armed forces once they "have fallen into the power of the enemy." (Janis and Noyes 2006, 530–31, quoting, inter alia, Dormann 2003, 45–46)

Supporters of the president's power to order the use of extraordinary rendition classify "terrorists" as outside the protections of the conventions.

Other treaties also make illegal the practice of extraordinary rendition. The 1966 International Covenant on Civil and Political Rights states: "No one shall be subjected to torture or to cruel, inhuman or degrading treatment or punishment"(Article 7). The 1951 Convention relating to the Status of Refugees can also be cited as affording protection against torture and refoulement of individuals "with a well-founded fear of persecution" (Article 1(A)(2)) (All Party Parliamentary Group on Extraordinary Rendition 2005, 9).

There are, as well, international norms that represent shared understandings on the part of nations regarding acceptable standards of behavior. The obligation regarding non-refoulement prohibits states to send individuals to countries where they face prosecution or torture, although under the norm there is an exclusion for some types of terrorist criminals.

The United States has consistently maintained that human rights treaties are limited in their application only to American soil, and in this regard, the Bush administration insisted that Guantánamo detainees are exempt from the mandates of domestic and international law because where they were being held lay outside the domestic domain. Guantánamo, however, has special territorial status, which makes it analogous to territory of the United States.

Furthermore, the argument has been made that customary law bears directly on the legal legitimacy of extraordinary rendition. Customary law is seen as the norms and rules customarily followed by civilized nations and binding on the states. Since many countries besides the United States cooperated in extraordinary rendition by providing facilities, aiding in transportation, and cooperating in interrogation with the use of disputed techniques, extraordinary rendition can be said, untenable as it seems to us, to be in line with customary law.

Another possible source of legal interpretation of extraordinary rendition is the International Criminal Court and the criminal tribunals with special jurisdiction in Rwanda and the former Yugoslavia. The Parliamentary Group of the United Kingdom's House of Commons believed that these forums are relevant to extraordinary rendition. The treaties on which the courts are established dictate that supposed or denominated military actions can be tried as criminal acts. "Reflecting the seriousness of the offense of torture an evolving body of international law also requires criminalization and prosecution of ancillary acts, such as complicity to, and aiding and abetting, torture" (All Party Parliamentary Group 2005, 11). Criminalization of complicity is not unusual in international law.

The nature of the American legal system constitutes a vital element in discussions of the legality and morality of extraordinary rendition. American jurisprudence is marked by advocacy and is concerned with statutes that frequently are, and sometimes deliberately so, ambiguous. There often is no undisputed single meaning of the words written by people, often groups of people, who are seeking a degree of consensus. The enactment also can contain a high level of abstraction. This situation allowed President George W. Bush, when discussing the legality of tactics employed in extraordinary rendition, to state: "Because of the Supreme Court's ruling that said we must conduct ourselves under Common Article II of the Geneva Convention says that, you know, there will be no outrages upon human dignity. It's like—it's very vague. What does that mean, 'outrages upon human dignity'?" (*New York Times* 2006)

DOMESTIC LAW IN THE UNITED STATES

In the United States, there are a number of official and semi-official statements that attempt to answer the question: What is torture? With regard to Article I of the CAT, the United States, while adopting a definition generally compatible with the CAT language, has also made the distinction that "an act must be specifically intended to inflict severe physical or mental pain or suffering."

Relevant domestic law includes both statutory provisions dealing with rendition and torture and constitutional law regarding the authority of the executive.

As to rendition, the authority to send a detainee to another country or jurisdiction was enunciated by Chief Justice Charles Evans Hughes in the Supreme Court decision in *Valentine v. United States.* Hughes ruled that "the power to provide for extradition is not confided to the Executive in the absence of treaty or legislative provision." Thomas Jefferson, when secretary of state, advised the president: "The laws of the United States, like those of England, receive every fugitive, and no authority has been given to their Executives to deliver them up" (Moore 1891, 23). In his treatise on extradition, John Bassett Moore, summarizing precedents, concluded that "the general opinion has been, and practice has been in accordance with it, that in the absence of a convention or a legislative provision, there is no authority vested in any department of the government to seize a fugitive criminal and surrender him to a foreign power" (1891, 21).

As to torture, the language in the US Torture Victim Protection Act is straightforward (28 U.S. C. Section 1350). Section 2(a) defines torture as "any act ... by which severe pain or suffering ... whether physical or mental, is intentionally inflicted on that individual for such purposes as obtaining from the individual or a third person information or a confession."[8]

During the Bush administration, several promulgated opinions addressed elements of extraordinary rendition. These so-called torture memoranda defined torture as "equivalent in intensity to the pain accompanying serious physical injury, such as organ failure, impairment of bodily function, or even death" (Bybee 2002, 1). The memoranda concluded that the Geneva Conventions do not apply to suspected terrorists.[9] Further, the memos suggested that the president could authorize violation of the Torture Statute when he believed it necessary in an emergency. The rationale offered was that "Congress may no more regulate the President's ability to detain and interrogate enemy combatants than it may regulate his ability to direct troop movements on the battlefield." For some, the constitutional analysis trumped all other considerations for determining the legality of extraordinary rendition and torture. They emphasized Article II of the Constitution that declared that "the President shall be Commander in Chief of the Army and Navy of the United States, and of the militia of the several states, when called into the actual service of the United States."

So: Is extraordinary rendition part of the power or other aspects of the executive's power? Some legal scholars answer with a perfervid "No":

> These renditions are accomplished outside of treaties and courts and are solely the result of presidential fiat. The claims to executive power to undertake these actions are of recent origin ... the great balance of U.S. history and law weighs against these claims. The actions of the Bush administration have been aided by judicial deference, a deference that challenges historical assumptions and law concerning executive power. (Weaver and Pallitto 2006, 102)

Other legal scholars concur in this judgment:

> [I]n a democratic republic the most defensible version of commander in chief is a separationist conception that is far narrower than the Bush Administration believes. It was understood as a limited authority at the time of the constitutional framing, and the subsequent evolution of the U.S. military has not broadened it. (Luhan 2008, 484)

A contrary school of thought answers "Yes"—the president does enjoy the right to employ extraordinary rendition tactics. This position requires accepting the notion that when Congress authorized hostilities in Iraq, it implicitly authorized any means necessary in the view of the Commander in Chief to pursue those hostilities. There is a middle position as well that is labeled the Zone of Twilight. It maintains that both the president and the Congress have powers in the pursuit of war, and what the executive has done in regard to extraordinary rendition represents a plausible interpretation of his prerogatives. This argument includes a view that colloquially is labeled "Tit for Tat": the executive can retaliate for cruel and outrageous acts (e.g., 9/11 and al Qaeda).

But the question remains whether the Constitution is even relevant to rendition or, at the very least, whether it should prevail in circumstances as threatening as the post–September 11 period when equally heinous actions appeared to be imminent. The significant consideration here is whether the Constitution exists to protect the American people or to protect the country's institutions. If the former, should not the executive do what it takes, even if what is required is contrary to constitutional provisions, to protect the people? If the latter, then the loss of American lives may be a price the country must be willing to pay in certain circumstances for the upholding of constitutional principles.

A fundamental or first principle regarding the Constitution and presidential powers was enunciated by John Yoo (2004), an attorney in the Department of Justice, who maintained that "statutes and treaties must be interpreted so as to protect the President's constitutional powers from impermissible encroachment and thereby avoid any potential constitutional problems" (1230). Yoo concluded that the president cannot violate the Constitution when he is pursuing his war powers.

To the non-legally trained observer, the range of opinions bearing on the subject of executive authority may seem remarkable and confusing—a verbal quagmire. Supreme Court Justice Robert Jackson, who had served on the court in the Nuremberg trials of charged axis war criminals, summarized the issue in *Youngstown Sheet & Tube Co. v. Sawyer* with words that provide a firmer context for the array of views on the subject under review:

> [One] may be surprised at the poverty of really useful and unambiguous authority applicable to concrete problems of executive power as they actually present themselves. Just what our forefathers did envision, or would have envisioned had they foreseen modern conditions, must be divined from materials almost as enigmatic as the dreams Joseph was called upon to interpret for Pharaoh. A century and a half of partisan debate and scholarly speculation yields no net result but only supplies more or less apt quotations from respected sources on each side of any question. They largely cancel each other.[10]

Thus, while we conclude that the jurisprudential arguments against the legality of extraordinary rendition are persuasive, we cannot say that those that reach an opposing position are so weak as to be indefensible.

BEYOND LEGAL ANALYSIS

Some persons believe that it is preferable to override legal analysis and concentrate on the morality of extraordinary rendition; others are primarily interested in its effectiveness.

In regard to morality, some scholars and ethicists have focused on means-end criteria. The ends of the practice can be viewed as essential and admirable; they are variously articulated as defeating al Qaeda, winning the war against terrorism, or making the world safer for democratic institutions. The means that may be employed to reach these ends generally are passed over lightly, if considered at all. Such ends are considered to justify the deeds deemed necessary for their accomplishment, even if some means are at best questionable.

The moral dimension of the issue is often addressed by the construction of hypothetical situations. "Ticking bomb" scenarios are usually introduced to address the morality of using an extreme measure to stop a particularly horrible outcome. The scenarios portrayed include a nuclear bomb set to destroy a major metropolitan area, or a bomb soon to be detonated on a jet, the identity of which only the person being interrogated knows. The best response to the vignettes used to defend torture, and one with which we agree, has been expressed by Christopher Kutz (2007), a professor at the law school of the University of California, Berkeley:

> At the level of philosophical reflection, the ticking bomb example does show something. It shows that we can imagine limits to even our most deeply held moral principles. But we should use this realization to strengthen our principles and their application in the world, not to abridge them. Here is how the realization of imagined limits can strengthen principles such as the right against torture. By their very divergence from real situations (existential necessity can exist only in a hypothetical world), imagined scenarios like the ticking-bomb example can continually remind us that we have not reached the imagined limits in reality, that we need to push our principles further and ever further, that if we relinquish our deepest precepts in an ideal world of imagined scenarios, there will surely be no hope for them in the real world we all inhabit. (276)

Kutz has elaborated elsewhere on this fundamental issue. Again we concur:

> I have argued that the attempt to justify both torture and extra-legal authority by reference to necessity fails. It rests upon a conflation of necessity as fact with necessity as justification and on a broad and deep misunderstanding of the nature of pre-institutional rights. The bedrock principles we have, concerning the dignity of humanity and the limits of legitimate power, are hard-won achievements of the last several centuries. Scrabbled together out of convention, claimed in the shadow of authoritarian power, they have become the marks by which we know our moral

> identities as both persons and nations. Threats and emergencies demand response, but that response must be grounded in a confidence in our principles' abilities to meet the demands of the world on our own terms. This confidence is equally a form of judgment, the determination that threats to our interests not be confused with threats to our existence. Far more dangerous to us, to who we are, is the threat of finding necessity in every conflict with evil and emergency in every war. (275)

Ethical stances on extraordinary rendition, like the laws and treaties, vary in their conclusions. A sample is found in statements during the 2008 US presidential campaign regarding harsh interrogation techniques, practices that have come to be euphemistically camouflaged as "enhanced interrogation." Rudolph Giuliani urged "people who do the interrogation to use every method they could think of." John McCain wondered, "How in the world do we condone such things?" Mitt Romney concluded that it was wise "to leave it somewhat unclear." Hillary Clinton concluded that "the decision to depart from standard international practices must be made by the president." John Edwards predicted that "saying no to torture will protect our troops and our values," and the emergent victor, Barack Obama, stated categorically, "No administration should allow the use of torture."

The effectiveness of extraordinary rendition poses further interpretative problems. Does the practice work? Does it achieve the ends desired? Do the interrogation methods elicit important intelligence from subjects, and do they save innocent people and inculpate terrorists? Are the methods essential to reach that goal?

There is no dominant scientific consensus on these questions. One cannot conduct controlled experiments, although there are some laboratory simulations that have a sliver of relevance. But what evidence we have is primarily anecdotal.

The chief of the CIA's bin Laden unit during the Clinton administration characterized extraordinary rendition as "the single most effective counterterrorism operation ever conducted by the United States Government" (Scheuer 2007, 14). Others have maintained that torture administered in front of family members is particularly effective. President Bush, in vetoing a bill to ban waterboarding, called tough interrogation methods "one of the most valuable tools in the war on terror." He asserted that enhanced interrogation methods used against terrorism captives had helped prevent several disasters, from a planned strike on the US consulate in Karachi, Pakistan, to a plot to hijack a plane and fly it into Los Angeles's Liberty Tower (now the US Bank Tower), the city's highest structure. Allegedly, al Qaeda had recruited South East Asians for the deed on the ground that they would arouse less suspicion than Arabs. The president said that because the Army Field Manual is a public document and available on the Internet, his administration thought it prudent to develop alternate techniques to glean information from prisoners (Serrano 2008).

Former vice president Dick Cheney, as well as President Obama's director of national intelligence and four successive CIA directors also claim that harsh interrogation methods produced valuable information. Cheney was particularly forceful in staking that claim: "They did work. They kept us safe for seven years" (Shane 2009a). The inference rests upon questionable logic, particularly since reliable documentation of this claim has not yet been forthcoming on the ground and that to do so would compromise state secrets.

Others offer very different assessments of the efficacy of torture. Few can doubt that in some instances accurate data are secured, but the question remains as to whether such results could not otherwise have been obtained and whether the process generates a great deal of misleading information. Malcolm Nance (2007), who trained American military personnel to resist torture, described one of his experiences: "On a Mekong River trip [in Vietnam] I met a man.... In torture, he confessed to being a hermaphrodite, a CIA spy, a Buddhist Monk, a Catholic Bishop, and the son of the King of Cambodia. He was actually just a schoolteacher whose crime was that he once spoke French."

The creator of the CIA's extraordinary rendition program has observed that information that is secured usually is "very tainted" by foreign agencies that deal with suspects at the request of the United States. Some veteran trainers in the SERE (for Survival, Evasion, Resistance, and Escape) program seeking to prepare potential American captives to resist the effects of torture concluded that physical pressure is less reliable than other interrogation tactics. Government studies conducted in 1950 found the Chinese communist interrogation techniques that employ "crude tactics," such as forced standing for hours, "had produced false confessions from captured American pilots." Under such abusive treatment a prisoner became "malleable and suggestible, and in some instances he may confabulate" (Shane and Mazzetti 2009, A14). A government memorandum, "Operational Issues Pertaining to the Use of Physical/Psychological Coercion in Interrogation," concluded, "the prisoner's physical response to the pain inflicted by an interrogator would obliterate such nuance and deprive the interrogator of these key tools [such as rapport with the prisoner].... The key operational deficit related to the use of torture is its impact on the reliability and accuracy of information provided" (Joint Personnel Recovery Agency 2009, 2).

An FBI agent who questioned Abu Zubaydah, allegedly a chief recruiter for al Qaeda, wrote, "There was no actionable intelligence gained from using enhanced interrogation techniques on Abu Zubaydah that wasn't, or couldn't have been gained from regular tactics. In addition, I saw that using these alternative methods on other terrorists backfired on more than a few occasions—all of which are still classified" (Soufan 2009).

Very strong evidence about the unreliability of information elicited by torture can be found in the annals of witchcraft prosecutions on the European continent. Hundreds, perhaps thousands, of persons confessed after being tortured and were burned for acts that we can now appreciate they could not conceivably have done, such as flying through the air to diabolic conventicles.

Finally, the effectiveness of interrogation procedures that do not resort to torture can be considered. "Techniques based on earning the interrogated person's trust, even affection, learning about his or her family, developing a sense of liking and trust ... may elicit important intelligence" (Kleinman and Alexander 2009). It has been reported that the most valuable information that Abu Zubaydah gave up resulted from a "traditional, rapport-building approach led by two FBI agents" and that waterboarding and other subsequent torture did not produce useful intelligence (Shane 2009b).

One also needs to address the extent to which the international reputation of the United States is damaged by public revelations of episodes of extraordinary rendition. If that reputation becomes increasingly soiled, this may encourage the recruitment of terrorists and the performance of terrorist acts against Americans and American institutions.

The once rare practice of extraordinary rendition is a kind of Rorschach test that indicates how one views the goals and functions of government. The authorized use of kidnapping on foreign sovereign territory and the torture of those kidnapped forces us to confront the fundamental moral and legal issue of why law and government exist, as well as the hierarchy of principles that should guide state action. Along the way, it brings into play questions concerning necessity, provides a context for consideration of national values, and raises important questions about priorities and means-end considerations.

For those like us who believe that the practice of extraordinary rendition must be halted, several routes to that end are available. They include domestic laws challenging the notion of an all-powerful commander in chief, actions to erode the use of the state secrets doctrine, legislation clarifying the scope of the nation's compliance with non-refoulement norms, and statutes tightening the definition of torture. In addition, opponents of the practice have access to universal criminal jurisdiction in some nation states, which allows national courts to assume jurisdiction for crimes committed anywhere, resort to actions in the International Criminal Court or the International Court of Justice, and campaigns to embarrass offending nations. Civil suits against third parties involved in extraordinary rendition can be launched, and use can be made of domestic law on kidnapping, assault, conspiracy, and aiding and abetting. Habeas corpus suits filed in US courts offer another possible remedy.[11] The law may in places be ambiguous; it is not, particularly when it is aligned with morality, impotent.

NOTES

1. The authors wish to thank the UCI Interdisciplinary Center for the Scientific Study of Ethics and Morality; the Center for Global Peace and Conflict Studies; the Center for Law, Society and Culture; and the University of Milano, Bicocca, for their support of this work. Valuable assistance in research was provided by Brian Williams, UCI Library, and Angie Middleton, UCI Law.

This information became known in an Italian trial of the police officers, and in absentia, the CIA officials. In July 2005, Italian prosecutor Armando Spataro issued warrants for thirteen American CIA operatives (Tribunale della Liberta). Later, nine more Americans were added. In the trial, elements of the rendition were described. Marco Sabatino, a policeman and computer scientist, explained how the Italians had identified the kidnappers. The police obtained the telephone traffic records of the mobile phones that, on the day of the kidnapping, made or received calls within the "radio base cells" situated around via Guerzoni. Seventeen mobile phones were considered suspect: near the area of the base cell, they made numerous short calls to one another, mainly between 12:15 and 12:42 p.m., the estimated time of the kidnapping. The phone cards were activated or sold between November 2002 and January 2003; they ceased to be used two or three days after the kidnapping. Some of the cards were activated anonymously or without the holder's knowledge. After the kidnapping, the users of four of the cards went to the military base Aviano with the users of five other mobile phones picked up in the later investigations. The users followed the "Milano-Portogruaro" motorway and exited at Portogruaro around 4:00 p.m. and arrived about 4:30 p.m. at Aviano, in the radio base cell where the US Air Force Airport is located. While traveling to Aviano, they used the mobile phones to call the CIA Station Chief of Milan (Robert Lady), Aviano airport, and telephone numbers in Virginia (the home of the CIA headquarters). To identify those present at the crime scene, several other sources of evidence were presented, including photocopies of identity papers (passports and drivers' licenses) used in the hotels. (Personal communication from Armando Spataro, Milano, June 26, 2008.)

2. See generally Wells 2008.

3. Subsequently, a lawsuit on these facts was filed in Germany, and in April 2008, the American Civil Liberties Union filed a petition with the Inter-American Commission on Human Rights on behalf of El-Masri.

4. The Arar case is discussed in Carter (2005), Goodman and Goodman (2006), Mayer (2008), and by Arar's wife, Monia Mazigh (2008).

5. Under the Barack Obama administration with a reconfigured Congress, the early record was mixed. There continue to be bills introduced to address and clearly outlaw the practice: prohibiting the expulsion, return, or extradition of persons by the United States to countries engaging in torture, requiring the State Department to produce annually a list of countries where torture is known to occur, and prohibiting the reliance on diplomatic assurances of suspect countries. However, the administration invoked the state secrets doctrine in situations of alleged extraordinary rendition in the very first months of the administration.

6. John Bellinger, chief legal advisor to the Department of State, 2006, has said, "So we think that Article 3 of the CAT … is legally binding upon us with respect to transfers of anyone from the United States but we don't think it is legally binding outside of the United States" (quoted in Council of Europe 2008, 104).

7. But for some, the question is not difficult: "Unless you have been strapped down to the board, have endured the agonizing feeling of the water overpowering your gag reflex, and then feel your throat open and allow pint after pint of water to involuntarily fill your lungs, you will not know the meaning of the word … a controlled drowning.… Waterboarding is slow motion suffocation with enough time to contemplate the inevitability of blackout and expiration—usually the person goes into hysterics on the board" (Nance 2007). Whether waterboarding is torture was seemingly resolved under the administration of President Barack Obama, but its decision did leave open the possibility that under some conditions, techniques not in the Army Field Manual could be used.

8. Appended as statutory notes to the Alien Tort Claim Act, a federal law that states, "The district courts shall have original jurisdiction of any civil action by an alien for a tort

only, committed in violation of the law of nations or a treaty of the United States." Recent interpretations allow for American leaders to be held responsible for human rights abuses committed as a result of their presence in a foreign country. As to sanctions: "An individual who, under actual or apparent authority, or color of law, of any foreign nation—(1) subjects an individual to torture shall, in a civil action, be liable for damages to that individual."

9. The torture memos were rescinded in the waning months of the administration of George W. Bush.

10. Concurring opinion. See also Raven-Hansen (2008), quoting Bilder and Vagts (2004), who concluded that for issues of national security law and presidential power, like many issues of international law, "it is difficult to agree on any clear criteria or tests for determining when a legal argument or position ... is so clearly erroneous or politically slanted as to be simply 'out of the ball-park' and beyond the range of permissible good-faith argument" (693–94).

11. For a full list, see Weissbrodt and Bergquist (2006).

REFERENCES

All Party Parliamentary Group on Extraordinary Rendition. 2005. "Briefing: Torture by Proxy: International Law Applicable to 'Extraordinary Renditions.'" December 2005. http://www.chrgj.org/docs/APPG-NYU%20Briefing%20Paper.pdf.

Arar v. Ashcroft, 414 F. Supp. 2d 250 (E.D. N.Y. 2006).

Bilder, Richard B., and Detleve F. Vagts. 2004. "Speaking Law to Power: Lawyers and Torture." *American Journal of International Law* 98: 689–95.

Bybee, Jay S. 2002. "Memorandum for Alberto R. Gonzales Counsel to the President Re: Standards of Conduct for Interrogation Under 18 U.S.C. §§2340–2340A." August 1. https://news.findlaw.com/wp/docs/doj/bybee80102mem.pdf.

Carter, Jimmy. 2005. *Our Endangered Values: America's Moral Crisis.* New York: Simon & Schuster.

Council of Europe. 2008. *CIA above the Law? Secret Detentions and Unlawful Inter-State Transfers of Detainees in Europe.* Strasbourg, France: Council of Europe Publishing.

DiMento, Joseph F. C., and Gilbert Geis. 2006. "The Extraordinary Condition of Extraordinary Rendition." *War Crimes, Genocide & Crimes against Humanity* 2: 35–64.

Dormann, Knut. 2003. "The Legal Situation of 'Unlawful/Unprivileged Combatants.'" *International Review of the Red Cross* 85: 45–74.

El-Masri, Khaled. 2005. "Statement: Khaled El-Masri." *ACLU,* December 6. http://www.aclu.org/human-rights_national-security/statement-khaled-el-masri. *El-Masri v. United States,* 479 F. 3d 296 (4th Cir. 2007).

Goodman, Amy, and David Goodman. 2006. *Static: Government Liars, Media Cheerleaders, and the People Who Fight Back.* New York: Hyperion.

Janis, Mark W., and John Noyes. 2006. *International Law: Cases and Commentary,* 3rd ed. Thompson/West.

Joint Personnel Recovery Agency. 2009. "Operational Issues Pertaining to the Use of Physical/Psychological Coercion in Interrogation." Accessed April 25, 2009. http://www.washingtonpost.com/wp-srv/nation/pdf/JPRA-Memo_042409.pdf.

Kleinman, Steven, and Matthew Alexander. 2009. "Try a Little Tenderness." *New York Times,* March 11.

Kutz, Christopher. 2007. "Torture, Necessity, and Existential Politics." *California Law Review* 95: 235–76.

Luhan, David. 2008. "On the Commander in Chief Power." *Southern California Law Review* 81: 477–570.

Mayer, Jane. 2008. *The Dark Side: The Inside Story of How the War on Terror Turned Into a War on American Ideals.* New York: Doubleday.

Mazigh, Monia. 2008. *Hope and Despair: My Struggle to Free My Husband, Maher Arar.* Toronto, Canada: McClellan & Stewart.

Mohammed et al. v. Jeppesen Dataplan, Inc. 579 F. 3d 943 (9th Cir. 2009).

Moore, John B. 1891. *A Treatise on Extradition and Interstate Rendition.* Boston: The Boston Book Company.

Nance, Malcolm. 2007. "Waterboarding is Torture … Period." *Small Wars Journal,* 31 October. http://smallwarsjournal.com/blog/2007/10/print/waterboarding-is-torture-period.

New York Times. 2006. "President Bush's News Conference." September 15.

Raven-Hansen, Peter. 2008. "Review Essay: The Terror Presidency: Law and Judgment Inside the Bush Administration, by Jack Goldsmith." *American Journal of International Law* 102: 672–84.

Scheuer, Michael F. 2007. "Statement of Mr. Michael F. Scheuer, Former Chief, Bin Laden Unit, Central Intelligence Agency." Extraordinary Rendition in U.S. Counterterrorism Policy: The Impact on Transatlantic Relations. Joint Hearing before the Subcommittee on International Organizations, Human Rights, and Oversight and the Subcommittee on Europe of the Committee of Foreign Affairs. April 17. http://www.fas.org/irp/congress/2007_hr/rendition.pdf.

Serrano, Richard A. 2008. "Bush Vetoes Bill to Ban Waterboarding." *Los Angeles Times,* March 9.

Shane, Scott. 2009a. "Interrogations' Effectiveness May Prove Elusive." *New York Times,* April 23.

———. 2009b. "Divisions Arose on Rough Tactics for Qaeda Figure." *New York Times,* April 17.

Shane, Scott, and Mark Mazzetti. 2009. "In Adopting Harsh Tactics, No Inquiry Into Past Use." *New York Times,* April 22.

Soufan, Ali. 2009. "My Tortured Decision." *New York Times,* April 23.

United Nations Commission on Human Rights. 2006. "Situation of Detainees at Guantanamo Bay." February 15, 2006. http://www.globalsecurity.org/security/ library/report/2006/guantanamo-detainees-report_un_060216.htm.

United States v. Reynolds, 345 U.S. 1 (1953). http://www.state.gov/g/drl/rls/hrrpt/2003/27926.htm.

US Department of State. 2003. *Country Reports on Human Rights Practices 2003: Egypt.*

Valentine v. United States, 299 U.S. 5 (1936).

Weaver, William G., and Robert M. Pallitto. 2006. "Extraordinary Rendition and Presidential Fiat." *Presidential Studies Quarterly* 36(1): 102–16.

Weissbrodt, David, and Amy Bergquist. 2006. "Extraordinary Rendition: A Human Rights Analysis." *Harvard Human Rights Journal* 19: 123–60.

Wells, Holly. 2008. "The State Secrets Privilege: Overuse Carries Unintended Consequences." *Arizona Law Review* 50: 967–98.

Yoo, John. 2004. "Transferring Terrorists." *Notre Dame Law Review* 79: 1183–1236.

Youngstown Sheet & Tube Co. v. Sawyer. 343 U.S. 579 (1952).

Chapter 9

Ethics and Individual Freedom

Kristen Renwick Monroe, Chloe Lampros-Monroe, and Peter Hawkins

Conversation

Chloe Lampros-Monroe (CLM): How did you become interested in the field of ethics in general, and the subject of gender equality in academia in particular?

Kristen Monroe (KM): I was thinking about this because I knew we were going to do this interview. It's always interesting when you yourself have to answer the same question you pose to others; you see it from a different perspective, which is actually an important theme in ethics—seeing the world from other people's perspectives. The idea is that if you can see the world from another's perspective, you'll have more empathy for them and presumably better understanding, so you'll treat them better.

Thinking about this question, I realized that I'm a person who believes very strongly in freedom—individual freedom. For me it's a kind of passion; I feel about freedom the same way that a lot of people feel about religion. I'm not a religious person, but I believe strongly that you should have individual freedom. When I say that, however, it doesn't mean what it means to a lot of people in this country today, which is that we should get government off our backs. I think government is a tool that can be used for good or bad, and we should use government to make a more compassionate world for people, particularly for those who aren't making it and need help.

CLM: Like FDR?

KM: Yes, very much like FDR; I'm very sympathetic with some of his programs. A lot of things follow naturally from my belief in freedom. One of the topics I've been interested in—partly because of my passion for individual freedom but also because I'm a woman—is equality for women. I think this is a very important goal. When I was your age [16], women didn't have careers in academia. There were a few but there weren't too many of them. I had not planned to become an academic; I fell in love with social science. I wanted to be a judge like my dad, and I was going to be the first woman on the Supreme Court. I was that ambitious. After going to college, I was getting ready to apply to law school, and I was going out with two guys—one was going to go to grad school and the other was already in law school—and I realized that the work that the guy in law school was doing wasn't very interesting to me. My dad always talked a lot about constitutional law issues, and those were what I thought lawyers dealt with. But my law-school boyfriend was studying torts and business law and corporations and secure transactions, and I thought, "I don't want to work for a big company. I just want to do constitutional issues." So I thought maybe if that's what lawyers do, and I slowly realized only a small percentage of lawyers do what I thought was interesting—human rights law or constitutional law—then I would take a master's degree before I went to law school, kind of on the theory that I needed a little more time before I got serious in life. Like a lot of people, I believed what everybody said, which was that you have to have a plan and figure out what you want to do with your life. You probably hear at your school that you should have a five-year plan and a ten-year plan. It was all very serious. And I did have a plan—I was a very self-serious young girl—and so I applied for a master's degree.

Now this was in 1967, which was during the Vietnam War, and they were still giving draft deferments for graduate school, so a lot of guys who would not ordinarily have gone to graduate school were applying that fall, and it was really tough to get in. One thing they told everybody was that you always apply for a PhD, even if you don't want to get one. I'm really good at missing the forest for the trees—I can miss one tiny little fact, which changes everything—so I forgot that little bit of advice. I'm also very honest and it's really hard for me to lie, and I didn't want to get a PhD—I just wanted a master's degree—so that's what I applied for. I applied at only three places, all excellent schools on the East Coast. Then a friend of my father's said I should apply to the University of Chicago. I'm embarrassed to admit that I had all the snobbery that only a Midwesterner can have for the East Coast—I thought that East Coast was better than the Midwest, so I figured if it's in Chicago, how good can it be? But I applied and I got in. More important, I did not get in anywhere else.

I suppose in retrospect that this is an interesting story in and of itself. I think I didn't get in anywhere else partly because I came back to Smith College from my junior year in Geneva and was slow to find a good adviser. I was doing a thesis in international law and the professor who taught international law had been snapped up by everybody else in the Geneva group, who were all interested in international politics and who were mostly better organized than I was. So by the time I got around to asking Peter Rowe if he'd be my supervisor, he was over-booked. But I still needed an adviser. I was doing a thesis on how international organizations give force to their decisions, and I was looking at the case at the International Court of Justice that was deciding the South-West African Mandate question. South-West Africa (now called Namibia) had been owned by Germany, which then lost it during the First World War, and it was administered by South Africa, which was then a colony of Britain. The League of Nations was happy to have Britain administer it. But then when South Africa instituted apartheid, the international community was not happy to have them administer South-West Africa. So there was a question as to who was in charge of the Mandate territory. The woman supervising me wanted me to look up obscure stuff on Bantu tribes, and I wasn't interested in that; I wanted to deal with international law. So she gave me a B-plus for my working grade for my thesis, which was a bad grade for an honor's student, and I so didn't get into the other schools I'd applied for—places like the Fletcher School of Law and Diplomacy, where I now realize I would have been miserable; I would have dropped out. I probably would have gotten a master's degree. But they essentially train people for the Foreign Service, and I'm not that "establishment" of a person—I thought I was at the time, but in fact I'm really not. So this was an instance of something that at the time felt like a bad thing, but in retrospect, it was one of the best things that ever happened to me because I loved Chicago, and I can see now that it was probably the very best grad school possible for me.

So I ended up going to the University of Chicago and I was really happy there. But "The Plan" was still to take a master's degree and then go to law school. I'd also gotten more serious about this guy I was going out with, the academic guy; he was in grad school at Harvard and was putting a lot of pressure on me to transfer there or to the law school in Boston and get married. It was one of these things—your whole life, women of my generation were taught that the best thing for you is to marry a bright young man who's on his way up, and this guy was obviously a bright young man on his way up. We're all affected by culture, and I thought I should keep "The Plan," too. I thought I wanted to marry the bright young man.

That all came apart my first year in grad school. Up until then, I'd had no identity crisis at all. But that first year in grad school was miserable, probably one of the more miserable years of my life. But it really taught me the difference between the things that you think you want, the things you think you *should* want, the things you think other people want you to want, and the things you *really* want. Because I was so happy intellectually. I'd found a home, if you will. Up until then, I didn't know what social science was—nobody really told me about it. At Smith, at least in the 1960s, they tended to do very traditional political and legal histories, institutional forms of government and so on—they weren't doing social science yet. Being at Chicago was different, like somebody had pulled the blinders off, and I was just dazzled by everything I saw. Here I was, I'd been living in the slums of Chicago, spending all my time going to the library, and I was ecstatically happy intellectually.

My mother came up one time and said, "What do you do on weekends?" I said, "Well, I go to the library." She said, "What else do you do?" and I said, "I go to the library. . . ." She said, "What about for fun?" I said, "I go to the library!" She said, "What about the Art Institute and the Symphony?" I said, "Is there a Symphony in Chicago?" Now Chicago has one of the world's best symphonies. I didn't know it. She said, "Get dressed, we're going to the Art Institute!" So she took me downtown, and I did enjoy the Art Institute. But when she left, I just went back to the library! To me it was like I'd found home—I was just really happy there. To this day if I'm unhappy, I go to the library and I'm very happy—it's safe and it's secure and the books are friends. It's as if you can pick a book off a shelf and have Aristotle talking to you, or Talleyrand or whomever you want. They're there for you. So I don't think books are impersonal; I think they're really alive and wonderful, and I'm just thoroughly happy going to the library.

So all the things that I'd thought I wanted—to be on the Supreme Court, to be a lawyer like my father, to marry the bright young man from Harvard who'd have a successful career—it turns out that when I had the chance to do them, I didn't want any of those things. So there I was. I didn't know where I was going. Everybody said you have to have a plan, and I didn't have a plan anymore. I just knew that I was totally, completely, utterly happy where I was at the moment. They were going to pay me to read these books, when I would have paid *them* a lot of money to read them. I had no idea where I was going, but I was *not* leaving. So I stayed and became a social scientist, and I've never looked back. It's been wonderful for me and a great life, and I've been very happy. It's about your passion—it's something that makes me feel alive to think about ideas. That's what I really like. My books and my kids—those are the two big things in my life. So I think you should follow whatever it is that works for you. I think you have to figure it out for yourself; no one else can do it for you. It's often close to what you think it is but significantly different, and you have to trust your instincts there.

So I've always been interested in what it is that makes people feel alive, and I think that has to do with individual freedom—for me it's all wrapped up in the same thing. So I was going to be in academia, and I didn't know what it meant to get a job. I knew that it was probably going to be hard for women, but I just knew I had to do it; I had to try. Initially, I got interested in very mainstream research. My thesis was on presidential popularity and the economy. I did some articles and a book on that topic, which is actually one of the all-time great books for curing insomnia—if you ever have any trouble sleeping, I can give you a copy, and it'll put you out right away. It was a solid piece of academic work but pretty boring. I'm certainly not especially proud of it, but I'm not *not* proud of it. I think what happens is that sometimes you learn to do something and it doesn't come out too great, but the fact that you've stretched means that the next thing you do might be better. That's really important; I always tell kids that. Don't be afraid to stretch and fail.

So when I'd done this book and stepped back to put it into perspective, it looked like a lot of the arguments in the field were actually about different approaches to the field. It was as if scholars were like the blind men looking at different parts of the elephant.[1] One person felt at the tail and described the elephant as rope-like, and another person felt the trunk and described the elephant as like a huge hose, or the leg felt like the elephant was a pillar, and so on. Then we are all arguing because the parts are different and each person thought the part they felt was the "real" elephant.

So I started to drop back and try to focus on a larger picture. It seemed to me that one of the most important theories in social science was the assumption that underlies a lot of disciplines (primarily economics, but also what are called rational choice models in social science), which is that individuals do best what's best for them, subject to their information and opportunity costs. For example, rational choice theory suggests if I think that I want to buy this kind of car, I go out and look and make a decision. And I thought, "Is that really the way people operate?" I thought that certain behaviors didn't fit into that model, and one of them was collective behavior, which is what people do when they're involved in politics—they organize interest groups or they vote and get together. There's no market mechanism. For instance, private citizens working in groups to save local land from development, giving their time to a cause—they're not going to get a directly proportionate benefit, but they're doing it for the greater good. So I wanted to look at collective action.

The other type of behavior that didn't fit into the assumption that individuals do what's best for them was altruism. I like to think of myself as an efficient person, so I thought, "Altruism's smaller. There's less of it to explain than there is collective action. I'll look at that first and clean it up, then turn to the big stuff." That was in 1988, and I'm still looking at altruism. I can only hope that all my students and my children find something like that—it was a research project that took me places I didn't know existed intellectually.

So I started getting involved in research on altruism. I read a lot of the literature on altruism. I just took two years off and read; I did no writing during that time. By the time it was over, I was pretty sure that I understood at least what the traditional wisdom said. I was pretty sure that there's a kind of Gemeinschaft/Gesellschaft distinction that people talk about. *Gemeinschaft und Gesellschaft* ("Community and Society") is a book by Ferdinand Tönnies (1957), which says that there are two different types of societies into which people are born. One is an individualistic society. This is over-simplifying a lot, but this view is very close to what we think of as the Anglo-American tradition. The other is more a community with community ties, such as peasant societies in East Asia, where you do things more for the good of the society rather than the individual. I figured altruists would be communitarians, people who thought in terms of the community, not individual self-interest. So I decided I'd design a project, and this is why I like doing empirical work because—maybe it's just that my imagination is limited, but in this case I think the imaginations of everyone else in the field were also limited—it turned out not to be that way at all. In fact, the people who are altruists are not community people. When I later did work on Nazis, it turns out the community people are much more likely to be the genocidalists because everybody who's in the community gets treated well, but if you're outside, you can get treated quite poorly. So it was something entirely different than what my reading had suggested. This is one reason I love empirical work. It forces you to construct systematic tests of ideas that you often take for granted, and you learn something new as a result.

I got involved in this project. I designed a continuum, so you would have at one end people who were more self-interested, such as entrepreneurs, then people who were philanthropists, who gave away some of their money but kept large chunks of it, and then people who had risked their lives for other people or who had gotten the Carnegie Hero Commission Award—people who jump into lakes or go into burning buildings to save people, and they're just kind of ordinary people, not professional policemen or fire-fighters or anything like that, and then the last group I interviewed were people who rescued Jews. I found that altruists were people who had a particular perception of themselves that meant they saw something in the other person—they felt that they were connected to other people through a common humanity. And that was nothing that I'd found in the literature I'd read.

There were a couple of important things that happened to me. I started out with the survey because I was trained as a behavioral social scientist, where you were supposed to survey the literature and come up with certain hypotheses, then test those hypotheses in a certain way. For instance, economists say that altruism is really reciprocal altruism—I'll do something nice for you so that you do something nice for me later. So I'd have a question, and I would ask people who were certified as being philanthropists or Carnegie Hero Commission Award winners, "Did you do what you did because you thought they [the person to whom they acted altruistically] would be nice to you afterwards?" So I had a whole series of questions, but when I started interviewing people, I realized it was really uncomfortable going in and saying, "Let me just start asking you some questions." So I said, "Let's just get to know each other—tell me a little bit about yourself," more because I was polite than for any thought-out methodological reason. But what that did was open a whole new area for me because people would come up with things that I would never have thought of in a million years. And what they self-selected was much more interesting, and I think much truer to what was really going on inside their heads than if I'd gone in with a pre-designed survey. I actually did ask all of the survey questions at some point, but I ended up doing what are called narrative interviews. At that point in time, people weren't doing those in political science, so I had to feel my way along and kind of invent a lot of it, and I had to go into other fields and do a lot of reading, which was good for me. So I was always kind of pushing and going into other areas that I didn't really know anything about. To do that, you just have to be willing to make a fool of yourself and say, "I don't know anything in this area, so tell me about it." People look at you like, "You don't know the most basic things about linguistics?" No, I don't. I don't know anything about linguistics. But they'll teach you or give you material to read, and you can learn it that way.

I found out that the psychological variables were the more important ones, so I ended up studying a lot of social psychology. After I sent the altruism book off, I kind of moped around the house for a week because I realized I really missed the people in the book.[2] They were gone. Then I thought that I could do some more interviews with the rescuers. I had used the first book to see what the altruists told me about social theory and political theory. I thought, "Well, this is moral behavior, so I'll go back and look at the interviews and read material in moral philosophy and learn about that." I'd never studied any philosophy, so it was all new to me. I had to learn it from scratch. I was very lucky to be made a Laurance Rockefeller Fellow at Princeton, so I had a chance to work with some of the philosophers there, and I learned a lot about philosophical theories. So I went through and tested some of those theories using my interview data. But here again, what I found is that they also didn't work. My data were closer to virtue ethics, but that theory didn't entirely explain what I had found empirically. What I found was more character and identity. But it wasn't *just* character; it was how you saw yourself, how you saw your character in relation to other people. So you could be a really lousy human being, but if you saw yourself as a good person, then maybe you'd respond that way. Or you could be a really good person but not see yourself that way. Or maybe you were a bad person—like Oskar Schindler probably was from an objective point of view—but you could still do a good thing because you saw yourself in a certain way in relation to a certain person or group of people, as Schindler saw himself as responsible for some reason to the Jews he saved. So it was the perceptions that were important, and that was really interesting to me.

That experience reawakened my interest in ethics, and I realized then that this had always underlain my interest in politics because I think politics is the study of power and influence, but it also has a large component that's normative. Politics is not just about describing the world as it is; you have to ask yourself, "Is this the way the world *should* be?" Because, going back to what I said earlier, I think politics and government should be a tool to make life better. It shouldn't just be something to protect you from each other. It should do that, of course, but it also should be a tool that helps you care about your fellow human beings because the world is an unfair place, and some people systematically get left at the bottom. Those people can be certain ethnic groups, economic classes, or racial groups. When certain people consistently get put at the bottom, government needs to step in to try to minimize that—because I think every person deserves the chance to lead a flourishing life. I don't think you deserve having it given to you, but I think you deserve a chance to try to find it for yourself, and if there is systematic discrimination against one group, then government needs to step in. Government has to be one area where the playing field is a little more equal than it is ordinarily in life.

So this all leads to my work in the chapter in this book—my work on gender equality. I noticed that there are a lot of inequalities in academia. It *should* be that things are being corrected because since the sixties, we've been concerned with these issues for women, and there have been a lot of government policies that are supposed to open things up for women. One of the arguments that was often given was that it's a pipeline problem; that in order to have a full professor, you have to have someone trained who's competent—you don't want to put a woman in who isn't qualified to teach, or do heart surgery, or whatever it is they're doing—so you have to start with the earlier time periods and open up opportunities for women, or minorities, at that level, at the entrance level. You want female full professors? Then train more female graduate students. That made a lot of sense, except that after forty years, what do the statistics say now on women in academia? I looked at those with a student [named] William Chiu, and it seemed that overall, the statistics haven't gotten better—they have improved a little bit, but not as much as you would have hoped, given the number of women going into graduate school. So what we did in the chapter was to go through and look at data from the American Association of University Professors, and it turns out that now there are slightly more women going to graduate school than men. But once women start in grad school, their percentages start dropping down; when you get to PhDs, it's less than 50 percent for women; then for hiring assistant professors, it's less again and beyond that it keeps dropping down. So that to me suggests that you still need institutional policies to ensure gender equality. Just making sure you have more women coming along isn't going to help. The so-called pipeline leaks and it leaks women.

I think there are two main factors here that may explain the leaks in the pipeline and why it is women who are leaking out of the route toward top academic positions. One is that women have families, and I think women still spend more time with the family. Things have gotten better, but all the evidence I've seen is that women still do more housework and take on more of the childcare and caring for elderly relatives and so on. So you need to have a different kind of model for professional success. We need a model that's not a straight linear model. A linear model would be one where I get my PhD, then I get my first job, then publish, and so on. It all follows in a straight line with little deviation. But women may want to have a child at some point and may want to take four, five, or even ten years off, and they should be allowed to go back in without being penalized as long as they can keep up in their field. In some fields, it's very difficult to do, and I personally am very discouraged since I don't know of any solution for situations where women have to be in a laboratory or a clinic or so on—a doctor has to be in her office, a lawyer has to be in court. It was much easier for me as an academic—I worked late after the kids went to bed, and I can work at home a lot of the time. I could take books and get some work done in the car while I was waiting to pick the children up from somewhere. I'd usually have an article I was reading or writing or a transcript that I was editing, and I'd just fill it in with my spare time here or there. I can't do that if I'm a brain surgeon—I can't just have a client come along with me for me to operate on while I'm waiting in the car to pick up my child from school!

So I think that's one of the reasons why it's more difficult for women in the bench sciences, and why you've seen such little progress. But I also think there's a lot of prejudice there. So the article that William and I did was designed to look at the prejudice issue. It's a difficult issue that we're trying to look at because sometimes when you talk about it, people feel accused—a lot of men sometimes feel put on the spot. I should be clear. I don't think that existing discrimination is necessarily the result of overt discrimination so much as insensitivity on people's parts and institutional structures that are outdated. American colleges have a tenure policy where you have to have tenure at a certain time, and we need to modify this that so it's more flexible. So I think there are a lot of things that you can do to change the situation—and that was one of the things we were talking about in chapter 9. But you also have to have good data to tell you what the situation is before you can talk about how you change it.

So that's how I got interested in the work I did for this chapter and I think it illustrates how using tools of science can help us analyze issues—such as gender equality—that touch on ethics in a way that reveals things we might not otherwise find.

NOTES

1. The famous Indian parable illustrating how each of us is limited by our own perspectives and perceptions.
2. Monroe 1996.

REFERENCES

Monroe, Kristen. 1996. *The Heart of Altruism: Perceptions of a Common Humanity.* Princeton, NJ: Princeton University Press.

Tönnies, Ferdinand. 1957. *Community and Society (Gemeinschaft und Gesellschaft).* Translated by Charles P. Loomis. East Lansing: Michigan State University Press.

Gender Equality and the Pipeline Problem in Academia

Kristen Renwick Monroe and William Chiu

Investigation

Gender equality in American society remains a controversial topic. In a year that saw the political candidacies of Hillary Clinton and Sarah Palin founder amid charges of differential standards for women, charges that themselves then became contentious, it is clear that gender remains a salient political issue. Recent scholarly studies have suggested that the problem of gender discrimination extends beyond politics and the workplace, reaching into academia itself. Studies using aggregate data find both wage and hiring differentials (NSF Science and Engineering Indicators 2008) and a glass ceiling (Cotter et al. 2001). Scholars also find surprising gender bias in more focused assessments of worth, such as the differential evaluations of academic citations (Johnson 1997) and the peer-review process (Wenneras and Wold 1997). Extensive in-depth interview data with female faculty at both the Massachusetts Institute of Technology (MIT 2002, 1999) and the University of California at Irvine (UCI; Monroe et al. 2008) offer more detailed systematic evidence that gender equality in academia remains only a partially realized goal, with women far too often struggling to crack a glass ceiling that remains strong and oppressive. The implication of these studies is that more active policies by university administrations are required to eliminate gender equality in academia.

The "pipeline" solution is frequently offered as one response to such reports of continuing gender inequality, in academia as in other workplace situations. The pipeline solution suggests that problems such as tokenism and status expectations will be ameliorated naturally if more women enter the pool of potential faculty or, in the world outside academia, the pool of potential managers and top administrators.

As the number of women increases, sheer numbers and forced social contact should mitigate stratification based on homophily (preference for same gender), segregated social networking and mentoring, and the tendency to characterize success and authority in masculine terms (Roth 2004). Logically, one can imagine that a rich pipeline of qualified individuals will ultimately result in women rising to the top over time, thus thawing the glass ceiling naturally and without requiring specific policies to end gender discrimination.

Unfortunately, studies of academic and commensurate private sector and federal jobs indicate that a larger pipeline does not lead to more women in higher-status positions, given the problems in the advancement mechanism (Myers and Turner 2004; Powell and Butterfield 1997; Roth 2004). If advancement is improved by mentoring possibilities, the findings that the lack of women in positions to mentor produces inferior results when compared with men (Smeby 2000) offer further discouragement. The evidence on the pipeline as a solution to gender inequality thus suggests a closer look is in order before accepting this as a viable solution to the problem of ongoing gender discrimination in academia.

In this chapter, we thus examine arguments concerning a pipeline problem to ask whether gender discrimination in academia is being corrected as more women enter the graduate school pool and move through the system. Our analysis uses aggregate data, collected by the American Association of University Professors (AAUP), on the status of women in higher education in the United States over (roughly) the last thirty years of increased efforts to increase the pool of females entering the academic market. Analysis of these data suggests that the percentage of women in the Academy remains disproportionately low and that those women who do succeed in finding an academic job still—overall—earn less money for the same job as do male counterparts. The pattern in academia thus reflects the male/female wage gap that has characterized the US economy since analysts began tracking such data in the 1970s.

The question then becomes: Why does this pattern exist? Is it ongoing discrimination? Part 1 of our analysis outlines the legal and economic context for our analysis. In particular, we ask if there are explanations (Becker 1971) other than discrimination that might account for differential situations of male and female faculty. Part 2 analyzes these economic explanations using AAUP data and finds that no economic explanation can fully explain the gender gap. The persistence of ongoing, if subtle, gender discrimination remains the more plausible explanation. An aggregate analysis of statistical data on gender discrimination within academia thus confirms the more impressionistic interview data presented in the UCI and the MIT studies. Glass ceilings are multiple and resilient. Worse, analysis suggests gender inequality will not be alleviated significantly through the natural projection of more women into the pool of graduate school. More positive efforts will be necessary to end gender inequality in academia.

BEYOND ECONOMIC LOGIC AND OSTENSIVE TITLE IX COMPLIANCE

The university offers an ideal target to explore the gender equity gap in depth. The product of the university is knowledge creation distributed in two primary forms: the education of young adults and the advancement of the knowledge base. Assuming men and women possess similar capacities for intellectual work, biased distribution along gender lines should not emerge. Social structure also may play a role. Private actors driven by profit are typically organized as top-down hierarchies in which bosses make decisions that employees follow. In contrast, universities are collegial in organization, incorporate vote-based decision making, and—at least theoretically by opening space for deliberation—accommodate bottom-up input. Intuitively, the university should distribute rewards equitably.

Statistics from the AAUP, however, suggest this is not the case. In academia as in the wider market, women earn approximately 19 percent less than men at all professional ranks (AAUP 2008). If economics and sociology fall short in explaining the gap, what else might account for this skewed distribution? Taking a moment to reflect on the bias in pay, if one changes frames from economic logic to a social psychological one, bias transforms from a systematic coefficient into a human propensity. As a form of prejudice or discrimination, bias produces differential outcomes. Glass ceiling and channeling effects, due in part to the tenure and advancement structure of the university, interact with family demands that attach disproportionately to women and that suggest a discouraging intractability for gender differentials.

To explore these possibilities, we can examine the economic institutional reality of a Research I university—such as UCI or MIT—and examine these data in conjunction with the findings from the qualitative data on attitudes and evaluations on the part of faculty toward work conditions. Inflection points illuminated by the interview data in the UCI and the MIT studies thus focus attention on areas where glass ceiling and funnel effects may occur. Macro-level data reveal in broad strokes the inequities attaching to women in the university.

In reviewing the following data, one should be mindful of an important externality that attaches to public universities. Under Title IX, the Education Amendments of 1972, "no person in the United States shall, on the basis of sex, be excluded from participation in, be denied the benefits of, or be subjected to discrimination under any education program or activity receiving Federal financial assistance" at educational institutions, including most universities, save for a few, explicitly delineated exceptions (Title 20 U.S.C. § 1681). If economic actors do not have an incentive to pay a premium for maleness per se (Becker 1971), and if the law embraces an equal-pay-for-equal-work meritocracy, why do women earn less regardless of rank? The simple, off-the-cuff answer is discrimination. The challenge is finding a way to elicit discrimination as a variable conducive to social scientific inquiry.

The virtue of legislation such as Title IX is in the mandate for equal treatment. The pitfall is that the reality of a law on the books may lead to assumed compliance. For a rational individual, the law is a clear standard; one would be foolish to discriminate, given the disincentive of potential punishment, and deviation is unlikely given the validity of the norms. Following such assumptions, one would not be able to explain a male/female wage gap. The problem and these norms can, however, be viewed from another angle. Perspective matters, and the logic inherent in a particular framework can blind analysts to phenomena and variables not accounted for by the paradigm in which the analyst works. An alternate account locates law as a line in the shifting sands of institutional and cultural practice. The black letter of Title IX demarcates acceptable practice with the caveat that transgressions must pass muster in a court of law. The reality is that the capacity to bring suit hinges on many factors, including documented evidence, resources to retain legal representation throughout a lengthy process, and the mental and emotional fortitude to pursue a path that may create difficulties in the workplace and inflict personal stress.

Conceptually, if a model construes law as a nominal declaration of a formal political expectation, separate from or above existing practice, it then can account for the gap between expected and actual behavior. The advantage to this view is that it embraces actuality and compares analytic standards, whether economic or legal, with real experience. When laws reverse accepted practice, one cannot expect a custom to vanish from people's behavior instantly. One important consequence may be that while the university may adopt an anti-discrimination policy, harmonization between regulations and the law is greater than that between regulations and the actions of people to whom the regulations apply.

The conundrum may be that a new public norm has taken root where, because discrimination has been disallowed, it must not occur. This notion hinges on the assumption, common to economics, that individuals are rational agents and will not act against their interests. As a result, assuming that discrimination is unlikely to occur, the burden of proving harm shifts to the victim as the discrimination goes underground. This is precisely why interview data matter. Multiple accounts may corroborate experience of women as a group. In addition, even if women who report discrimination comprise a small minority of faculty, their voices should be seen not as a function of their number but rather as a signal that amplifies an important social and legal violation. Tokenism and the scarcity of group members in a population may exacerbate division and the rigidity of boundaries when dominant members stereotype or segregate naturally and thus marginalize token members (Kanter 1977). In this sense, voice serves, given the magnitude of transgression, as the proverbial canary in the mineshaft, a sign that conditions for all are deteriorating and must be addressed. The UCI and the MIT studies suggest this is precisely what occurs in academia. Our aggregate analysis must take this into account.

WAGE AND RANK DISPARITIES ACROSS INSTITUTIONS

Is academia truly as bad as a mineshaft? The economic data provide a compact if troubling overview of the breadth and depth of the problem. The university is unique in that it is both the exclusive supplier and dominant employer of qualified individuals. Our study focuses on research universities requiring doctoral degrees: typically PhDs, JDs (law), or MDs (medicine). Since this chapter focuses on academics rather than professionals, we limit discussion to PhD degree holders; our preliminary work suggests similar findings for professional schools. The university is partitioned into professional groupings demarcated most significantly by the tenure condition. The nature of tenure matters, and organizational controls for tenure and promotion will be evaluated for their potential effect in creating a glass ceiling.

While the male/female wage gap has decreased from an estimated high of 36 percent in 1979 to approximately 20 percent in 2005 (Doms and Lewis 2007), much of the improvement occurred in the 1980s; the current disparity has persisted since 1994. For occupations that require brute strength, one can plausibly posit pay being linked to an output that depends on physical capacity. Given that women and men possess equal capacity for intellectual work, however, it is puzzling why pay for workers who trade in knowledge—as is the case in academia—should segregate along gender lines. Nonetheless, the phenomenon in higher education appears to occur on two levels—stratification by institution and by pay.

In academia, this imbalance manifests itself (1) in distribution of women by institution and job rank and (2) by pay at these levels. Understanding career progress requires a conceptual model. The glass ceiling is well-known in popular parlance but requires specification to deliver robust findings. If one faces a glass ceiling, progress upward is stilted, such that disparities worsen with seniority (Cotter et al. 2001). Returns to human capital decrease for women as a group; as a woman gains more skills, she is less likely, compared with her male counterpart, to garner equivalent benefits. Thus a glass ceiling in the Cotter et al. model is characterized by a steep slope. The challenge is to operationalize human capital beyond the common method of time on the job. Longevity in and of itself does not necessarily equate to performance; a mediocre worker may stay average his or her whole career. What matters is tangible achievement. In academia, one may find the answer to the differential rates of advancement by analyzing the ways in which promotion-relevant achievement is measured.

When analyzing job distribution, one can evaluate both the ratio of women to men and respective pay. These may be interrelated; a lack of women indicates the lack of qualified candidates while those who are hired are, within their pay grade, less qualified than their male counterparts. The first issue can be viewed as a pipeline problem where too few women earn advanced degrees or choose an academic career upon completion of degree. The problem then is supply, not hiring. What about pay differences? Women may receive less pay than men for a variety of personal reasons, ranging from less skill or qualification to having less human capital to offer or being less committed to a work schedule because they require flexible schedules including part-time work or scheduled time off. In this view, women receive what they deserve. On the other hand, discriminatory practices also may explain dislocations in supply and pay.

For example, the natural progression for the academic professional who achieves tenure normally requires about seven years after receiving the PhD. During this time, advancement and ultimate retention depend on performance evaluation.

Table 9-1 provides a snapshot of activity in American graduate schools. Note that the data are not longitudinal and do not show a particular cohort's performance; rather they provide an approximation of women entering and completing a doctoral program. Women comprise the majority of graduate students who were enrolled for the first time in 2006 in doctoral programs (Council of Graduate Schools 2007). For the same year, women comprise less than half (48 percent) of all doctoral degrees awarded. The rates of men and women sum to 100 percent, so for degree attainment, share garnered by one gender affects the performance of the other. The point to note here is that men show a markedly higher rate of degree completion compared to their enrollment. There are a variety of reasons why people do not complete degrees. The average time to complete a doctoral degree is long, and many people may change their goals because of time, financial expense, family, or personal life. It is also possible that women who do not complete their degrees have chosen family or other considerations. The expense in time and money has, for a variety of reasons, not convinced these women to accrue the costly and yet basic human capital required for an academic career. In this sense, one can reasonably hypothesize that those women who do graduate are fully committed to a career, such that fallout for family reasons should be mitigated.

The data raise two separate groups of questions. In absolute terms, the pipeline for women academic talent appears quite rich. Women comprise a strong majority of first-time graduate students, a statistic that bodes well for the future for those who care about gender equality. In addition, despite an apparent 6 percent attrition rate, women still manage to earn 48 percent of the doctorates awarded. With respect to employment, women constitute roughly half of the newly minted and available academic work force. The question thus becomes whether equality in achievement is reflected as equality in employment in both qualitative and monetary aspects.

Table 9-1 Graduate School Enrollment and Doctoral Degree Attainment 2006 (CGS 2007)

	Women Grad Enrollment (%)	*Men Grad Enrollment (%)*	*Women Degree Attained*	*Men Degree Attained*
Doctoral Programs	54%	46%	48%	52%

Each year, the AAUP surveys wages across higher-education institutions with jobs sorted by academic rank and institutional type. In 2007, for the first time, the AAUP reported a breakout of employment data by gender. The survey is comprehensive, with 1,386 institutions, covering 367,476 faculty, reporting rank and salary data. The magnitude of the economic benefits being distributed can be estimated with a simple calculation of average salary multiplied by personnel count. For 2007 to 2008, nearly $28 billion was spent on salaries, with an average additional 21 percent spent as part of total compensation, creating in wages alone a $35.5 billion industry.

The problem is clearer when viewed in a matrix where the three ranks of professor (assistant, associate, and full) are plotted against institutional type (see Table 9-2). Women, for example, comprise less than 36 percent of faculty in more prestigious, doctoral-granting institutions but about 50 percent at less prestigious two-year colleges. In addition, women earn between 7 to 10 percent less per job title and, given the disparity in total numbers, up to 20 percent less than men as a group. Though startling, the economic data indicated here do not comprise the whole picture. Indeed, while economic explanations excel at providing a topographical survey of structural conditions, they may not provide deep insight into the experiences and challenges that those who traverse this daunting terrain must face. The distribution both between and within institutional categories suggests a glass ceiling that shunts women away from research universities to schools with lower career ceilings and acts as a barrier to advancing to higher stages. To examine gender equity in academia, let us now chart the economic data to depict the conditions women face in higher education.

As shown in Table 9-2, women comprise two-fifths of all faculty, tenure and non-tenure. A quick glance shows the disparity in rank, with men outnumbering women at senior positions three to one. Indeed, given the weighted decision making because of seniority in promotion and advancement, one can see how structural characteristics favor men. At many universities, promotion and advancement depend on a vote in which franchise varies by rank. Only Academic Senate faculty may vote; this group typically includes tenured faculty and may in special cases include non-tenured lecturers with security of employment contracts. For a candidate under consideration for a given rank, only those who possess that rank may vote. Thus, while all tenured

Table 9-2 Men and Women Shares of Academic Ranks (AAUP 2008)

All Institutions	*Percentage of Faculty*	
Rank	*Men*	*Women*
Professor	23.5	8.2
Associate Professor	15.9	10.8
Assistant Professor	14.1	13.2
Lecturer	2.8	3.4
All	59.7	40.3

faculty may vote on an assistant professor up for tenure, only those who already are full professors may vote to elevate a candidate. If men dominate a given rank, and if men have an implicit preference for other men (or opposed to women), women may fail to advance into what has been noxiously termed an old boys' club. Under such voting procedures, one can imagine how the ranks of a male-dominated elite group will change slowly, if at all. This contrasts with a system where a chief executive can elevate a protégé—man or woman—despite the objections of others, since command is top-down and is focused rather than flat and distributed. On the other hand, it is possible that the deliberative nature of tenure and promotion decisions may allow minority voices to persuade the majority to not act on implicit attitude, offering relief which is not available in single decision-maker conditions.

Men continue to dominate the professorial rank. It may be that many earned their positions years ago, when the pipeline was predominantly producing male academics or when women were denied entry into the Academy (Romito and Volpato 2005). Given that tenure assures security of employment extending at least into one's sixties and that overall faculty size is sensitive to a budget based in part on enrollments that may be static, the proportion of male professors may not be indicative of active discrimination against women; the disproportionate dominance of male full professors simply may reflect a first-in/first-out principle, in which qualified women have to wait until positions become available. (This would correspond with the pipeline argument, in a slightly different form.)

To determine whether a glass-ceiling effect exists, one can examine the distribution of faculty in the promotion lifecycle, focusing especially on entry-level effects at the assistant professor level. A survey of all ranks shows that this ratio favors men, although the gap at the entry level, assistant professor, is narrower than in other ranks. Intuitively, one might interpret the trends and remark that in terms of the academic talent lifecycle, universities are closing the gender equity gap for recent hires. On the other hand, it may be that women are being channeled from one category of institution to another. Aggregate averages across institutional type may collapse data on the extremes to the center. In this case, women would be filling tenure-track positions at less prestigious higher-education institutions.

Table 9-3 describes the distribution of tenure and non-tenure faculty by gender. More than four-fifths of men have or will have job security compared with two-thirds of women. The disparity is especially notable for those with tenure, as men report a nearly 50 percent higher success rate than women. Let us again consider the multiple possible interpretations these data offer. Tenure differences may reflect seniority because of a pipeline that was once male-dominated. The fact that more than a quarter of women (as compared with a fifth of men) are on tenure-track may suggest that younger women are doing well. Or it may indicate that women are achieving tenure more slowly than men. Finally, the largest difference is in non-tenure track. More women are in non-tenure track jobs by a healthy margin (70 percent), indicating that women are not able or willing to gain tenure-type jobs.

Table 9-3 Percentage of Faculty by Track and Gender

	Percentage of Cohort	
Track	*Men (%)*	*Women (%)*
Tenured	60.0	41.6
Tenure-track	21.1	26.2
Non-tenure	18.9	32.2

Where one has tenure also makes a difference. A closer examination reveals that the aggregate data mask disparities in the placement of women. From a pay and prestige standpoint, doctoral-granting institutions (hereafter Research 1) rank first, followed by master's-granting (universities), baccalaureate-granting (colleges), and finally two-year colleges (community colleges). To focus on rank alone, thus, is misleading; if women are professor at the community college level but not at university or above institutions, systematic channeling still may be occurring. Tenure provides security at the cost of mobility as jobs are a unique combination of research focus, campus, and regional location. Most important, trajectories differ at each stage with research universities offering the greatest potential.

The data suggest women show up disproportionate to their share of PhD graduates at non-research institutions. Do women choose liberal arts colleges over research institutions, or do forces beyond their control push them in that direction? Here, interview data may prove illuminating. If job seeking is a decision dependent on multiple criteria, some of which can be measured objectively and others which must be elicited, the reports of women at various stages of their academic career may help explain the observed pattern of employment.

Table 9-4 shows the disparity in distribution with a trend that favors men holding true at all levels. This table shows women's differential return to work and underscores the misleading nature of the overall data specified in Table 9-2. The gap between men and women at research institutions is remarkable and troubling. At all levels of tenure-type positions, women are heavily out-numbered, with the skew greatest at the highest rank, where men exceed women by a nearly four to one ratio (28.1 percent vs. 7.2 percent). In addition, there are nearly as many male associate professors (16.4 percent) as there are total female tenured professors (9.8 percent women associate + 7.2 percent = 17 percent). Given that assistant professor is an entry-level position that should reflect trends for recent graduates, the 23 percent higher rate for men (13.4 percent to 10.9 percent) can be interpreted either as a sign of marked improvement relative to that for senior faculty or that bias persists despite economic or moral good sense since new entrants should face a level playing field.

The trend continues at universities and colleges, where men possess a more than two-to-one advantage at full professor and a 31 percent advantage at the associate professor level. Only at the assistant professor level do women achieve near parity. At the lowest level, women and men are evenly divided at each rank, although men

Table 9-4 Percentage of Faculty by Academic Rank at Major Institutional Classifications by Gender

	Doctoral (Research)		*Master's (Universities)*		*Bachelor (Colleges)*		*Two-Year (Community)*	
Rank	*Men*	*Women*	*Men*	*Women*	*Men*	*Women*	*Men*	*Women*
Professor	28.1	7.2	19.6	8.3	19.8	9.2	15.7	14.0
Associate Prof.	16.4	9.8	15.6	11.5	16.4	12.5	11.5	11.4
Assistant Prof.	13.4	10.9	14.7	15.1	15.6	16.5	12.5	14.4
Instructor	2.1	3.1	2.4	4.1	2.8	4.1	7.7	9.3
Lecturer	3.3	3.9	3.0	3.8	1.0	1.1	1.4	1.9
Other	0.9	1.0	0.9	0.9	0.5	0.5	0.1	0.2
Overall	64.2	35.9	56.2	43.7	56.1	43.9	48.9	51.2
Faculty Total Size	178,584		118,557		50,557		19,778	

nominally outnumber women professors (15.7 percent to 14.0 percent), while the reverse is true at the assistant level (14.4 percent women to 12.5 percent men). Relative gains matter, as does absolute distribution. Four-fifths of all ranked academic jobs are at research institutions (46.2 percent) and universities (30.7 percent); colleges account for 13.1 percent, community colleges a meager 5.1 percent, with the remainder of jobs at two-year colleges that do not rank faculty. Applying the ratios in Table 9-2 to the total faculty row at the bottom of the table returns absolute numbers that show the disparity between men and women faculty in a different light. Male associate professors at research institutions (29,330) nearly outnumber the total of all female professors (30,138). Equally striking, there are four male *assistant* professors at doctoral institutions for every five female professors at *all* levels. From an overall labor pool perspective, it is clear that distributions are completely skewed in ways that segmented relative gain data do not readily reveal.

What are the chances that a woman ends up at a lower-tier school? Table 9-5 shows the distribution of women and men in the Academy. More than half of all male professors find work at research institutions. In contrast, more than one in five women end up at a college or community college. Viewed from this perspective, it is clear that women are branching into lower-ranked institutions.

Table 9-5 Where Women and Men Are Employed (AAUP 2008)

Placement	*Percentage of Men*	*Percentage of Women*
Research	52.3	43.2
University	30.4	35.0
College	12.9	15.0
Two-Year College	4.4	6.8

Why are there more women at lesser institutions? Is it due to structural impediments? Choice? It is possible that the difference in demands between a publication-oriented research institution and a teaching-focused college attracts or shuns men and women at different rates, depending on how one views the job placement process. One question worth noting is whether a "publish-or-perish" norm, often given anecdotally as the life of an R1 institution, can be determined objectively. If so, then one can see how heavy research institutions attract those who look for a particular lifestyle. On the other hand, if tenure guidelines, in which four areas of accomplishment are considered for advancement (only one of which is publication) are to be accepted at face value, then the publish-or-perish paradigm operates as an informal rather than an explicit barrier.

The higher-education labor market is less liquid than the private-sector market in that jobs are attached to schools that are located in particular places with job openings for narrow skill sets. Moreover, if faculty size is generally static from year to year, jobs become available only when a current faculty member leaves or if the department is committed to expansion. If a person chooses a lower-level institution, given career inertia and the rate of human capital accumulation disparities (in quality and volume) between lower and higher levels, that person is committed to a career at a lower rung. That is, jobs become open in a particular field only every so often, and achievements at lower levels are discounted such that faculty at two-year institutions have a low likelihood of moving up in institutions as time goes on. Given the lower and flat trajectory, entry into the two-year college seems unappealing. Figure 9-1 graphically depicts the stark differences between career paths. Not only is the starting point for a woman entering academia poor at the community college level, but also income inequality increases with seniority.

For comparison, the trajectory for men at research institutions is also shown. The slope of the curve appears to match that of women, indicating that men and women garner wages at a constant difference throughout their lifetime. Money is not everything, but given the many years it takes to earn a PhD, one should not overlook the diminishing returns to time that occur when a woman works at a lesser institution.

Under a rational economic actor model, a cost-benefit analysis guides decision making. Hence, before we explore the psychological aspects of this situation, we carry forward the discussion of the personnel distribution to wage structure. What might women be foregoing, if disparities are indeed the result of choice rather than discrimination? Table 9-6 shows the average salary for men and women by rank and institution type. At research institutions, women's salaries track men's at an approximate 90 percent rate at each level. The gap appears to narrow as one moves to lower status institutions, with women earning nearly the same as men at the community college. On the other hand, the skew in number of faculty reprises itself here, depressing the overall average. As seen in the last row of Table 9-6, once aggregated for each institution, the mean wage drops precipitously, with women

Figure 9-1 Average Salaries by Gender, Rank, and Institution Type

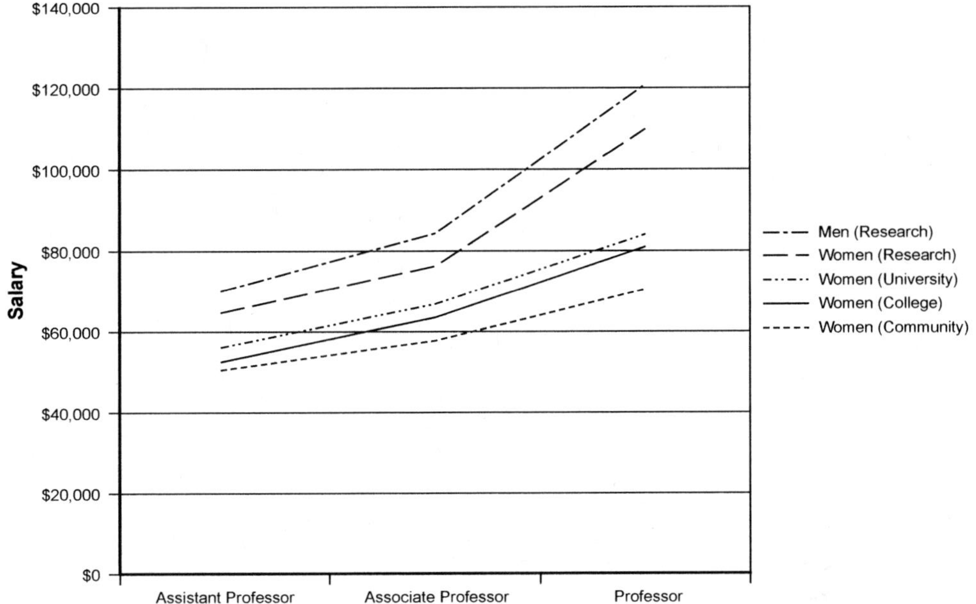

now earning 5 to 21 percent less than men as a group. Recalling the overall market of about $28 billion, the share women earn as a group in absolute terms is $5.6 billion less than men.

Perusing Table 9-6 reveals that the overall share as a percentage of men's salary reported in the last row is low and non-intuitive when viewed in the context of the detailed data. Perhaps the counts of male and female professors at different ranks account for the depressed aggregate mean wages. Comparing average wages by gender as a ranked interval of all wages sheds light on why the overall averages are so depressed. Table 9-7 lists the distribution of men and women's pay based on the mean salary for a given rank against all salaries of that rank. (See Table 9-9 details on the actual salary amounts.) The average male professor's wage at

Table 9-6 Women's Average Salary as a Percentage of Men's by Rank and Institution

Rank	*Research*	*University*	*College*	*Community*
Professor	91.0%	94.9%	95.3%	96.4%
Associate Professor	92.3%	95.8%	97.8%	97.3%
Assistant Professor	92.0%	96.0%	97.6%	98.0%
Overall	78.2%	87.6%	89.8%	95.3%

doctoral- and master's-granting institutions ranks at approximately the 61st percentile. Conversely, women's professor salaries rank at about the 50th percentile. A significant gap holds true at all levels and at all institutions. Assuming that rank is a suitable proxy for work performed, Figure 9-1 reveals that equal work does not merit equal pay. If one were to collapse the two curves to a single graph, a bimodal distribution is likely to emerge, with the mean for women to the left (lower pay, same rank) of men.

If seniority and experience determine one's wages, the fact that academia has been dominated by men would show up as a disparity that widens by rank. A three-tier ranking provides insight into gross differences but may not adequately capture the granular differences between work and pay for men and women. For example, a person who has published more and has spent more time within a rank will earn more than someone who has just been promoted to that same rank. At the University of California, the largest single university system in the United States, advancement proceeds according to a published schedule. There are three levels of professor; assistant professor has six steps, associate professor has five steps, and full professor has nine steps, with evaluation and promotion to each next step occurring approximately every two to five years (see Table 9-9). For example, a professor at step nine earns about $133,500 while a professor at step one earns $70,200. If men were the first to become professors and remain active, mean wages would be weighted in favor of men due to the seniority-accomplishment factor and to preponderance of male faculty who earned tenure two or three decades ago. Another possibility is that men are employed in fields that have higher average wages. If men were employed at a higher rate in science, technology, engineering, and mathematics (STEM), average wages would reflect this distributional difference. The University of California, for example, provides different schedules for those in health sciences, law, and business/economics/engineering (see Table 9-8). While beyond the scope of this review, research controlling for academic field and publication rate in the humanities (Ginther and Hayes 2003) suggests that this issue is one worth exploring in greater depth.

Table 9-7 Percentile Scale of Average Salary by Gender, Rank, and Institution Type

	Percentile Rank							
	Doctoral		*Master's*		*Bachelor*		*Two Year*	
Rank	*Men*	*Women*	*Men*	*Women*	*Men*	*Women*	*Men*	*Women*
Professor	62	49	61	54	59	53	56	51
Associate Professor	67	51	60	49	57	51	53	46
Assistant Professor	66	53	66	61	60	52	50	50

Table 9-8. Average Salary by Field (UC 2008)

University of California System	*Professor-Step III*
Law	$95,000
Business/Economics/Engineering	$109,500
Health Sciences	$135,600

If disparities exist, the hope is that current and future generations suffer less or not at all. This is where a focus on the assistant professor level becomes germane. If the process in the past was not merit-based, but the process today is merit-based, the return to human capital would be gender-blind and pay would be equal. Given that rank tracks time at some level, one can imagine that wage disparities worsen at more senior levels. Table 9-6 shows that at the doctoral institution, a woman earns about nine-tenths of what a man earns at all levels. The data used in this study do not allow for greater insight as to the causes of the persistent 8 to 9 percent gap between men and women. It is possible that women enter tenure track one step behind men, due to externalities that limit the accumulation of human capital. For example, if a woman were to start a family in the dissertation or post-dissertation phase, even a period of six months to one year could adversely impact her ability to publish or present her work and enhance her vita, thus leading to a starting point in the tenure process that remains one lock-step behind a man who has been free to focus solely on career. Note though that a woman who does not have children should have the same opportunities as a man, and thus the real driver of wage disparity under this view is family rather than gender. The decision to start a family appears to markedly affect women rather than men at the critical juncture in a nascent academic career (Mason and Goulden 2004). In a publish-or-perish advancement scheme, the lack of publications affects one's career (Long 1992). Following Erosa, Fuster, and Restuccia (2005), the interview data in the UCI and the MIT studies explored the importance of children as it relates to pursuit of an academic career and found this factor was relevant but that discrimination still existed, regardless of whether women had children or not.

Tables 9-6 and 9-7 also may be interpreted jointly. As one moves to lesser institutions, women's pay increases as a portion of a man's salary and the mean wage for each group tends to converge. Thus women gain relatively if they choose to go to a bachelor's-granting institution or, especially, a two-year college. Two questions arise: Why would pay go up for women as institutional prestige diminishes? Why would there be more women at lesser institutions as a proportion of overall faculty? Recall that at the doctoral level, women comprise about one-third of faculty and slightly more than one-half at the two-year level. Melinda Pitts (2002) argues that work decisions and occupational choice depend on human capital and on the perceived costs and expected utility of entering what she terms a female-dominated or non-female-dominated industry. Assuming comparable industries exist where one can choose between a similar job in either type of industry,

> an individual will only enter a female-dominated occupation if there is a net gain from doing so. If the wage an individual could expect to earn is lower in female-dominated occupations relative to non-female-dominated occupations, then the individual would only enter a female-dominated occupation if the penalty for doing so is less than the value of the nonwage characteristics of the female-dominated occupation. (Pitts 2002, 12)

For Pitts, an industry is female dominated when 60 percent of the workers, as measured by the Department of Labor, are women. In education, one might extrapolate this model and define elementary school as female-dominated with gradually decreasing shares of women workers as education level increases and with high school and two-year colleges comprised of a small majority of women culminating in a strong male majority by the doctoral level. To make sense of the female-dominated model for academia, one must also account for theories of human capital accumulation. Under the human capital model, a person will commit to acquiring more human capital, marketable skills that a person can offer on the job market, if the reward is worth the time and cost. For academia, there is a clear separation in minimum human capital at the elementary, high school, and college levels. Colleges, especially master's and doctoral institutions, require doctoral-level degrees, which take significant time and money to earn. For this reason, one cannot posit occupational choice in education as selecting from a continuum from elementary to doctoral, even if the participation rates appear to offer a coherent trend. Can the female-dominated model be used to parse higher education alone? In this view, women would choose between institutional types, where the two-year college is more female-receptive than the doctoral institution. What would the calculation for such a choice be?

On one hand, taking a job at a two-year college is likely less competitive and a woman can anticipate working with an equal share of women while earning about the same as a man. This appears then to be an equitable solution. On the other hand,

Table 9-9 Table of Average Salaries by Gender, Rank, and Institution Type ($)

	Research		*Universities*		*Colleges*		*Community*	
Rank	*Men*	*Women*	*Men*	*Women*	*Men*	*Women*	*Men*	*Women*
Professor	$120,661	$109,853	$88,632	$84,073	$84,829	$80,822	$73,024	$70,386
Associate Professor	$84,356	$76,155	$69,890	$63,465	$64,896	$63,465	$59,282	$57,692
Assistant Professor	$70,650	$65,002	$58,762	$56,402	$54,031	$52,710	$51,742	$50,697
Overall	$93,869	$73,383	$70,976	$62,153	$67,521	$60,631	$59,044	$56,298

foregoing a job at a higher-level institution returns tangible losses: women full professors barely earn more at two-year colleges ($70,386; see Table 9-9) than they do as assistant professors at doctoral institutions ($65,002). Moreover, the future income stream for a tenured professor at a doctoral institution is at least 56 percent higher ($109,853), a difference that totals more than $1.2 million over an eighteen-year career at the professor level after compounding at a modest 6 percent nominal interest rate. Including income differences in the prior two stages, one can imagine that if women are in fact choosing to participate in higher education at the bachelor and two-year college level, the female-dominated model would state that the non-wage characteristics of the female-dominated segment outweigh the uncertainty of the very large earning potential of higher-level institutions. Alternately, women might be participating at these levels due instead to structural impediments. Discriminatory practices or biases that disfavor the kinds of human capital women accumulate may lead to channeling women from research institutions to teaching institutions. None of this, of course, addresses the non-pecuniary values involved in teaching at the different type of institution.

The aggregate statistics confirm the discouraging view of gender equality gleaned from interview data about the situation for women in academia. National statistics from the United States make clear that there are fewer women working in academia than would be expected, given the relatively comparable numbers of graduate students coming on the market now. Women are employed at lower-status institutions. Women are ranked at lower grades and they earn less at each grade compared to their male counterparts. We have tried to scrupulously and objectively examine possible explanations other than discrimination but find these explanations unconvincing and unconfirmed by the aggregate data. Why would a woman fail to enter the Academy after earning a costly doctoral degree? Under a human capital argument, it defies economic logic for a woman to expend the effort in time and money to earn a credential and then not seek full financial recompense. No rational person would work so hard for less gain; indeed, one would surmise that the commitment to an academic career is tested in the graduate school process and that those less committed will drop out. Survivors should thrive precisely because they have demonstrated deeper commitment. Yet women, who comprise nearly half of all doctoral degree conferees, suffer an attrition rate of nearly one in six and comprise only 40 percent of professors.

In Cotter et al.'s construction, the glass ceiling should manifest itself in academia as a gate to higher status (promotion) and as increasing inequality at higher levels. Our analysis (Table 9-4) finds women do in fact participate at deteriorating levels as ranks rise at colleges, universities, and research institutions. At the highest level—the Research-1 institutions—women constitute 10.9 percent of assistant professors but only 7.2 percent of full professors. This disparity in numbers between men and women is both striking and extremely troubling, since—as the aggregate data suggest—this disparity represents a problem of advancement and not an absence of candidates. Income inequality within an institutional class does not appear to worsen on a per capita basis; at a given level, the gap in pay for men and women appears constant. The curves in Figure 9-1 show that women and men professors at research institutions appear to be the same, with women consistently earning from 3 to 8 percent less than men.

On the other hand, a striking disparity in pay emerges when one compares institutional classes. As women move from lower to higher institutions, pay inequalities increase. A 3 percent average gap at the community-college level becomes an 8 percent gap at the Research-1 institution. Why might this occur? Understanding the job placement system in academia as a whole may shed light on the distribution of jobs in the industry. The labor market is segmented into two broad categories: those with and without long-term security. It is further stratified into many small niches (expertise in departments, fields, subfields, and specialties). The nature of teaching and research distinguishes the advancement process in academia from the private sector. Freedom to pursue and share knowledge is provided by a tenure system. This holds important consequences for employment, as it creates an early gate that may act as a glass filter at an initial rather than later career stage.

How does tenure enter the equation? To be offered a tenure-track job is to assume that as long as the candidate delivers on her expected promise, a lifetime sinecure will be extended. Tenure-track employees are not "at-will" and cannot be easily fired or furloughed. As an organization, a university can offer a mix of jobs that enable it to maintain labor flexibility; some candidates are offered tenure-track, others lectureship. This mix may manifest itself systemically as a filter system that channels women to various locations in the university hierarchy. If gates between institutional levels are one-way, a person who starts out at a lesser organization is likely to be confined to advancement at that level. In this vein, two indicators may show how the glass ceiling manifests itself in the Academy more clearly. The headcount data in Table 9-4 indicate a quota system: men are overrepresented and, as a group, earn the lion's share of available income. This can be seen in the individual-level data, where women earn 90 percent of the amount men earn on a per capita basis but collectively earn 21 percent less. With women facing gaps of 13 to 30 percent in employment rate (Table 9-4) at all levels above the community college, despite receiving approximately half of all doctoral degrees, it is clear that women are showing up in the workforce at rates that demand additional investigation.

Two issues emerge. Where women go at the critical juncture of institution type matters a great deal, and the differential curves that attach to each show the slope difference between a community college and research university level (Figure 9-1). Thus, the glass ceiling may be unexpectedly occurring at the point of entry. Women in economics, for example, are significantly less likely to earn tenure at their first academic job (Ginther and Kahn 2004). Slower launch to a career means a shallow trajectory that results in lesser compensation over a longer duration. Location of hiring also matters. Prestigious universities may provide improved conditions for publication, which in turn lead to greater material benefits. A move up or down in prestige appears to lead to a statistically significant increase or decline in productivity (Allison and Long 1990). Tenure differentiates academia from the private sector because tenure track requires a large commitment in time, and tenure status itself is not fungible since higher education is not an open labor market. To be offered a tenure-track job is to be invited into an association that leads to permanent member status, as long as conditions of promotion are met. The intuition is that because tenure-track jobs show little churn, advancement is expected. Given this expectation, terms of entry take on added importance. (Estimates from the private sector jobs suggest professional women lose an average of $1 million over a lifetime of employment, merely for failing to negotiate successfully the terms of their first [Babcock and Lavescher 2003].)

The nature of voting rules also may act as a filter: senior faculty promote junior faculty to their ranks. While this process is opaque, one can imagine that the reasoning behind elevation and compensation decisions may be affected by implicit or even explicit discrimination. The elevation of a candidate necessarily involves subjective as well as objective metrics. If tenured and senior male faculty interact with junior and scarce female colleagues, interaction may follow archetypes described by theories of tokenism (Kanter 1977) and a fundamental disconnect in lived experiences in the workplace (Hale 1999). For women with multiple roles—be they ethnic, gender, or familial—the proportion of women of multiple marginalities also affects perceptions and demands in the professional role (Turner 2002). The role of women in governance may affect the procedures instituted for advancement (Moore 1987). Whether systematic patterns that favor men over women guide advancement or individual consideration results in unintended collective gender bias cannot be known without further investigation. Given the confidentiality of tenure and promotion, we can only surmise as to the types of practices extant. But the aggregate statistics presented here suggest the glass ceiling in academia is not one barrier at the very top of academia but instead exists at many parts throughout the academic workplace.[1] Since the percentage of women in the pool of professors has increased since the 1980s without comparable improvement in the percentage of women in the professoriate, the pipeline will not fix this gender inequity. More focused and systematic policies are required.

NOTE

1. We find it debatable whether women choose a job based on estimating the level of female domination within a particular subsector. If long-term earning potential is depressed at a college or community college, on economic logic alone it appears unlikely that a woman would choose these sectors, even if they are paid at a rate equal to that of men. Ceteris paribus, it seems logical that a choice to take a job at a college or a university is as much an exogenous as an endogenous choice. If research universities set criteria that cannot easily be

achieved by women—as a function of having children or caring for elderly family members, for example—then structural impediments constrain the array of choices, and women are forced to select from a subset of men's choices. If women choose a community college because they care less about money and more about non-monetary compensation, then college jobs may prove appealing. But when one considers the advantages—in terms of higher pay, higher status, less time spent in the classroom, and the increase in individual freedom, all of these factors provided by a research university—it seems illogical that women deliberately choose community college positions for family reasons. This is not to deny that work conditions can be construed in more than monetary terms. For academics, the resource pool one has on tap, in terms of colleagues, students, graduate program, libraries, and facilities among other things, may be of equal or greater importance than a salary. On a basic level, the desire to work at the university level and above may simply be an aspiration to actualize one's intellectual ambition. The economic data reveal a system that distributes monetary rewards unevenly, but the loss to the academic community may be greater. As a teacher, icon, role model, and mentor, a professor offers guidance to future generations of workers and leaders, scholars, and professionals. Voice as a function of critical mass may thus be silenced if, at the highest levels, one group dominates over the other.

REFERENCES

Allison, Paul, and J. Scott Long. 1990. "Departmental Effects on Scientific Productivity." *American Sociological Review* 55: 469–78.

American Association of University Professors (AAUP). 2008. "The Annual Report on the Economic Status of the Profession, 2007–08." Accessed April 20, 2008. http://www.aaup.org/AAUP/comm/rep/ Z/ecstatreport2007-08/survey2007-08.htm.

Babcock, Linda, and Sara Lavescher. 2003. *Women Don't Ask: Negotiations and the Gender Divide.* Princeton, NJ: Princeton University Press.

Becker, Gary. 1971. *The Economics of Discrimination.* Chicago, IL: University of Chicago Press.

Cotter, David, J. Hermsen, S. Ovadia, and R. Vanneman. 2001. "The Glass Ceiling Effect." *Social Forces* 80: 655–82.

Council of Graduate Schools. 2007. "Graduate Enrollment and Degrees: 1996–2006." Retrieved April 20, 2008. http://www.cgsnet.org/portals/0/pdf/DataSources_2008_03.pdf.

Doms, Mark, and Ethan Lewis. 2007. "The Narrowing of the Male-Female Wage Gap." Federal Reserve Bank of San Francisco Economic Letter, June 29 2007.

Erosa, Andres, L. Fuster, and D. Restuccia. 2005. "A Quantitative Theory of the Gender Gap in Wages." Federal Reserve Bank of Richmond Working Paper Series, 9.

Ginther, Donna, and Kathy Hayes. 2003. "Gender Differences in Salary and Promotion for Faculty in the Humanities 1977–95." *Journal of Human Resources* 38: 34–73.

Ginther, Donna, and Shulamit Kahn. 2004. "Women in Economics: Moving Up or Falling Off the Academic Career Ladder?" *The Journal of Economic Perspectives* 18: 193–214.

Hale, Mary. 1999. "He Says, She Says: Gender and Worklife." *Public Administration Review* 59: 410–24.

Johnson, Dan. 1997. "Getting Noticed in Economics: The Determinants of Academic Citations." *The American Economist* 41: 43–52.

Kanter, Rosabeth. "Some Effects of Proportions on Group Life: Skewed Sex Ratios and Responses to Token Women." *American Journal of Sociology* 82(5): 965–90.

Long, J. Scott. 1992. "Measures of Sex Differences in Scientific Productivity." *Social Forces* 71: 159–78.

Mason, Mary Ann, and Marc Goulden. 2004. "Marriage and Baby Blues: Redefining Gender Equity in the Academy." *Annals of the American Academy of Political and Social Science* 596: 86–103.

Massachusetts Institute of Technology (MIT). 1999. "A Study on the Status of Women Faculty in Science at MIT." Accessed August 10, 2008. http://web.mit.edu/faculty/reports/sos.html.

———. 2002. "A Study on the Status of Women Faculty in Science at MIT: 2002 Update of 1999 Study." Accessed September 1, 2008. http://web.mit.edu/faculty/reports/sos.html.

Monroe, Kristen, Saba Ozyurt, Ted Wrigley, and Amy Alexander. 2008. "Gender Equality in Academia: Bad News from the Trenches and Some Possible Policy Solutions." *Perspectives on Politics* 6: 215–34.

Moore, Kathryn. 1987. "Women's Access and Opportunity in Higher Education: Toward the Twenty-First Century." *Comparative Education* 23: 23–34.

Myers, Samuel, and Caroline Turner. 2004. "The Effects of Ph.D. Supply on Minority Faculty Representation." *The American Economic Review* 94: 296–301.

National Science Foundation. Science and Engineering Indicators 2008. Accessed January 10, 2009. http://www.nsf.gov/statistics/seind08/tables.htm#ch2.

Pitts, Melinda. 2002. "Why Choose Women's Work if it Pays Less? A Structural Model of Occupational Choice." Federal Reserve Bank of Atlanta Working Paper Series, 30.

Powell, Gary and D. Anthony Butterfield. 1997. "Effect of Race on Promotions to Top Management in a Federal Department." *Academy of Management Journal,* 40: 112–28.

Romito, Patrizia, and Chiara Volpato. 2005. "Women Inside and Outside Academia: A Struggle to Access Knowledge, Legitimacy and Influence." *Social Science Information* 44: 41–62.

Roth, Silke. 2004. "Opportunities and Obstacles—Screening the EU Enlargement Process from a Gender Perspective." *Loyola University Chicago International Law Review* 2 (1): 117–27.

Smeby, Jens-Christian. 2000. "Same Gender Relationships in Graduate Supervision." *Higher Education* 40: 53–67.

Title IX, Education Amendments of 1972. Title 20 United States code §1681.

Turner, Caroline Sotello Viernes. 2002. "Women of Color in Academe: Living with Multiple Marginality." *The Journal of Higher Education* 73: 74–93.

United States Department of Labor. "Title IX, Education Amendments of 1972." Accessed April 28, 2008. http://www.dol.gov/oasam/regs/statutes/titleIX.htm.

Wenneras, Christine, and Agnes Wold. 1997. "Nepotism and Sexism in Peer-Review." *Nature* 387: 341–43.

Index

About the Contributors

Born in 1921, **Kenneth J. Arrow** is Professor of Economics and of Management Science and Engineering Emeritus at Stanford University. He graduated from the College of the City of New York in 1940 and received the degrees of MS (mathematics, 1941) and PhD (economics, 1951) from Columbia University. He served in the United States Army during World War II, retiring with the rank of Captain, and has been research associate at the Cowles Commission (now Foundation) for Research in Economics and on the faculties of the University of Chicago, Stanford University, and Harvard University. His fields of research, primarily in economic theory and, to some extent, operations research, have included social choice, general equilibrium, information economics, inventory policy, medical economics, the economics of uncertainty, and conditions for economic growth. He has been president or officer of a number of learned societies. He has received the John Bates Clark Medal of the American Economic Association, the von Neumann Medal, the Nobel Memorial Prize in Economic Science, and the National Medal of Science.

Francisco J. Ayala is University Professor and Donald Bren Professor of Biological Sciences and Professor of Philosophy at the University of California, Irvine. Dr. Ayala is a member of the National Academy of Sciences (NAS), a recipient of the 2001 National Medal of Science, and served as chair of the Authoring Committee of *Science, Evolution, and Creationism,* jointly published in 2008 by the NAS and the Institute of Medicine. Ayala has received numerous awards, including the 2010 Templeton Prize for exceptional contribution to affirming life's spiritual dimension and honorary degrees from universities in nine countries. He has been president and chairman of the board of the American Association for the Advancement of Science and president of Sigma Xi, the Scientific Research Society of the United States. Ayala has written numerous books and articles about the intersection of science and religion, including *Darwin's Gift to Science and Religion* (Joseph Henry Press 2007) and *Am I a Monkey?* (Johns Hopkins University Press 2010). He teaches classes in evolution, genetics, and the philosophy of biology.

Warren S. Brown is Professor of Psychology and Director of the Travis Research Institute at Fuller, a research neuropsychologist with over eighty published scientific publications, and co-author of three books: *Whatever Happened to the Soul?* with Nancey Murphy and H. Newton Malony; *Did My Neurons Make Me Do It?* with Nancey Murphy; and *Neuroscience, Psychology and Religion,* with Malcolm Jeeves.

William Chiu is a PhD candidate in political science at the University of California, Irvine, with an emphasis in political psychology and political communication. He is currently writing his dissertation on the effect of the Internet on public sphere discussion and its implication for normative democratic theory. His research examines the nature of digital mass communication flows as they relate to the economic marketplace and First Amendment assumptions and protections. William holds degrees from U.C. Berkeley and Georgetown University and spent a decade working for high technology software and hardware companies.

Joseph F. C. DiMento is Professor of Law and Planning and the Director of the Newkirk Center for Science and Society at University of California, Irvine. Professor DiMento writes about and teaches domestic and international environmental law, land use and administrative law and regulation, and urban and regional planning. He has been at the University of California, Irvine, for many years in the planning and law and society departments of the School of Social Ecology. In 2007, he became a founding faculty member of the University of California, Irvine, School of Law where he also teaches international legal analysis. His research also involves environmental and urban law history. He has most recently authored *The Global Environment and International Law* (University of Texas) and (with Pamela Doughman, eds.) *Climate Change: What It Means for You, Your Children, and Your Grandchildren* (MIT). When not at the University of California, Irvine, he has worked for the US Department of Justice, served in regional and local government, and was a Fulbright scholar and teacher abroad. He is a member of the California Bar.

Gilbert Geis is professor emeritus in the Department of Criminology, Law, and Society at the University of California, Irvine. He received his PhD in sociology from the University of Wisconsin. He is a former president of the American Society of Criminology and recipient of its Edwin H. Sutherland Award for outstanding research. He has received similar awards from the Association of Certified Fraud Examiners, the American Justice Institute, and the National Organization of Victim Assistance. His most recent book is *A Documentary History of White Collar and Corporate Crime* (2011).

Peter Hawkins graduated from Oxford in classics in 2007. He participated in the project on moral courage while visiting at the Ethics Center during the summer of 2010.

Jennifer L. Hochschild is Henry LaBarre Jayne Professor of Government at Harvard University with a joint appointment in the Department of African and African American Studies and a lectureship in the Harvard Kennedy School. She taught at Princeton University before moving to Harvard in 2000. Hochschild recently co-edited (with John Mollenkopf) *Bringing Outsiders In: Transatlantic Perspectives on Immigrant Political Incorporation* (Cornell University Press 2009), and recently co-authored (with Brenna Powell), "Racial Reorganization and the United States Census 1850–1930: Mulattoes, Half-Breeds, Mixed Parentage, Hindoos, and the Mexican Race" (*Studies in American Political Development* 2008). Current book projects include *Transforming the American Racial Order: Immigration, Multiracialism, DNA, and Cohort Change* (co-authored) and *Facts in Politics: What Do Citizens Know and What Difference Does It Make?* Hochschild was founding editor of *Perspectives on Politics,* vice-chair of the Board of Trustees of the Russell Sage Foundation, and program co-chair for the annual convention of the American Political Science Association. She is a member of the American Academy of Arts and Sciences.

Cheryl Koopman completed post-doctoral training in psychology and public policy and international affairs at Harvard University and Columbia University. She was Assistant Professor of Clinical Psychology at Columbia University. She is now Associate Research Professor of Psychiatry and Behavioral Sciences at Stanford University, where she teaches on traumatic stress, research methods and statistics, and preparing the dissertation. From 2008 to 2009, she served as president of the International Society of Political Psychology. Koopman has over two hundred publications focused on the psychological consequences of traumatic stress in the medical or political context and on evaluating the effects of interventions on improving health and quality of life, especially for underserved populations.

Nicholas Lampros was graduated Phi Beta Kappa magna sum laude from the University of California, Los Angeles in 2004. A major in English and creative writing, Lampros served as sports editor of the *Daily Bruin,* has published several short stories, and is working on the great American novel. He is currently a Jerome Tobis Fellow at the University of California, Irvine, Ethics Center.

Chloe Lampros-Monroe is a junior at University High School in Irvine, California. She participated in this project while a 2010 Francisco Ayala Scholar at the University of California, Irvine, Ethics Center, where she played a key role in a project interviewing ordinary—and not so ordinary—people about how they make moral choices. She is interested in a wide range of subjects, from math and science to history and dance.

Adam Martin is a PhD candidate in political science at the University of California, Irvine. His work focuses on the role of identity and its contributions to both pro-social and anti-social behaviors, including theoretical explorations of neuroscience, religion, moral sense theory, and victims' reactions to terrorism. He has published five articles on neuroscience and ethics and is currently writing a dissertation on the political psychology of forgiveness in relation to the terrorist attacks of September 11, 2001.

Rose McDermott is a professor of political science at Brown University and a 2010–2011 fellow at the Radcliffe Institute for Advanced Study at Harvard University. A 2008–2009 fellow at the Center for Advanced Studies in the Behavioral Sciences at Stanford University, McDermott received her PhD (political science) and MA (experimental social psychology) from Stanford. McDermott has taught at Cornell and the University of California, Santa Barbara, and has held fellowships at Harvard's Olin Institute for Strategic Studies and Harvard's Women and Public Policy Program. She is the author of three books, co-editor of two additional volumes, and author of over eighty academic articles across a wide variety of disciplines encompassing topics such as experimentation, emotion and decision making, and the biological bases of political behavior.

Kristen Renwick Monroe is the founder and director of the University of California, Irvine, Ethics Center. She has published thirteen books, including the award-winning *Heart of Altruism* and *The Hand of Compassion.* The 2010 recipient of the Paul Silverman Prize for Outstanding Work in Ethics and the 2011 Ernie and Lucha Vogel Moral Courage Speaker at Principia College, Monroe also received two lifetime achievement awards given by the American Political Science Association: its 2010 Ithiel deSola Pool Award and its 2010 Goodnow Award. She has served as the president of the International Society of Political Psychology and vice-president of the APSA. Her latest book, *Ethics in an Age of Terror and Genocide,* is being published by Princeton University Press in 2011. She is currently working on a popular introduction to ethics and on books on Jewish émigrés from the Third Reich, gender equality, and empathy and ethics.

Gregory R. Peterson is Associate Professor of Philosophy and Religion at South Dakota State University, where he currently serves as program coordinator. Dr. Peterson's primary area of research is in religion and science and ethical theory, with special attention devoted to the biological and cognitive sciences and their implications for religious and philosophical approaches to human nature. Author of over thirty articles on religion and science in books, encyclopedias, and journals, he has published the book *Minding God: Theology and Cognitive Science* (Fortress 2002) and is co-editor of the forthcoming *Routledge Companion to Religion and Science.*

Bridgette Portman is a PhD candidate in the political science program at the University of California, Irvine, with a concentration in political psychology. Her interests include political ideology and belief systems, international relations, and existential psychology, and particularly the intersection of these fields. She has published articles on the nature of political psychology and is currently working on a dissertation about the relationship between political ideology and attitudes toward death.

Kevin S. Reimer is Professor of Psychology at Azusa Pacific University. His research program considers moral identity through natural language processing. Co-authored books include *The Reciprocating Self* (InterVarsity 2005) and *A Peaceable Psychology* (Brazos 2009). He recently authored *Living L'Arche* (Continuum 2009) and co-edits the forthcoming *Virtue Ethics, Exemplarity, and Ultimate Value* (Routledge forthcoming). Reimer's current book project details spiritual identity in the *Saints of 405*. He grew up in the San Francisco Bay area.

Thomas Schelling is a member of the National Academy of Sciences, the Institute of Medicine, and the American Academy of Arts and Sciences. In 1991, he was president of the American Economic Association and now is a distinguished fellow of the AEA. He was awarded the Frank E. Seidman Distinguished Award in Political Economy and the National Academy of Sciences Award for Behavioral Research Relevant to the Prevention of Nuclear War. Schelling served in the Economic Cooperation Administration in Europe and worked in the US White House and Executive Office of the President. He worked at Yale University, the RAND Corporation, and was a member of the Department of Economics and the Center for International Affairs at Harvard University. Schelling taught for twenty years at the John F. Kennedy School of Government where he was the Lucius N. Littauer Professor of Political Economy before moving to Maryland's School of Public Policy. Schelling shared the Nobel Prize in Economics in 2005 with Robert Aumann for advancing our understanding of conflict and cooperation through game theory analysis.

Michael L. Spezio is Assistant Professor of Psychology and Neuroscience at Scripps College and Visiting Faculty in Social and Affective Neuroscience at the California Institute of Technology. He is a social neuroscientist, investigating the psychology and neuroscience of emotional communication, empathy, compassion, political decisions, and moral action and how contemplative practices influence these processes. He is a co-editor on the forthcoming *Routledge Companion to Religion and Science* and on *Virtue Ethics, Exemplarity, and Ultimate Value* (Routledge forthcoming).

James Van Slyke received his interdisciplinary PhD from Fuller Theological Seminary in philosophy of religion with a minor in cognitive science. He is a research assistant professor at the Travis Research Institute in the School of Psychology at Fuller Theological Seminary and an adjunct professor at Azusa Pacific University where he teaches in the philosophy and psychology departments.

Nicole Wernimont is a doctoral student in clinical psychology in the Pacific Graduate School of Psychology-Stanford Psy.D. Consortium. Her main academic interests are in the theoretical intersections between continental philosophy and clinical psychology with a related interest in philosophical psychology. Her clinical interests include rehabilitative work with survivors of political trauma and ethical cross-cultural practice.